Lecture Notes in Artificial Intelligence 5400

Edited by R. Goebel, J. Siekmann, and W. Wahlster

Subseries of Lecture Notes in Computer Science

Michael Biehl   Barbara Hammer
Michel Verleysen   Thomas Villmann (Eds.)

Similarity-Based Clustering

Recent Developments
and Biomedical Applications

 Springer

Series Editors

Randy Goebel, University of Alberta, Edmonton, Canada
Jörg Siekmann, University of Saarland, Saarbrücken, Germany
Wolfgang Wahlster, DFKI and University of Saarland, Saarbrücken, Germany

Volume Editors

Michael Biehl
University Groningen, Mathematics and Computing Science
Intelligent Systems Group, P.O. Box 407, 9700 AK Groningen, The Netherlands
E-mail: m.biehl@rug.nl

Barbara Hammer
Clausthal University of Technology, Department of Computer Science
38679 Clausthal-Zellerfeld, Germany
E-mail: hammer@in.tu-clausthal.de

Michel Verleysen
Université catholique de Louvain, Machine Learning Group, DICE
Place du Levant, 3-B-1348, Louvain-la-Neuve, Belgium
E-mail: michel.verleysen@uclouvain.be

Thomas Villmann
University of Applied Sciences Mittweida
Dep. of Mathematics/Physics/Computer Sciences
Technikumplatz 17, 09648 Mittweida, Germany
E-mail: thomas.villmann@hs-mittweida.de

Library of Congress Control Number: Applied for

CR Subject Classification (1998): H.3.3, I.5.3, I.5.4, J.3, F.1.1, I.2.6

LNCS Sublibrary: SL 7 – Artificial Intelligence

ISSN   0302-9743
ISBN-10   3-642-01804-1 Springer Berlin Heidelberg New York
ISBN-13   978-3-642-01804-6 Springer Berlin Heidelberg New York

springer.com

© Springer-Verlag Berlin Heidelberg 2009

Typesetting: Camera-ready by author, data conversion by Scientific Publishing Services, Chennai, India
Printed on acid-free paper   SPIN: 12654949   06/3180   5 4 3 2 1 0

A physicist once came to Dagstuhl
and thought: 'The castle is quite cool!'
'So, clearly', he stated,
'we should replicate it!
And learn from one instance the right rule.'

Preface

Similarity-based learning methods have a great potential as an intuitive and flexible toolbox for mining, visualization, and inspection of large data sets. They combine simple and human-understandable principles, such as distance-based classification, prototypes, or Hebbian learning, with a large variety of different, problem-adapted design choices, such as a data-optimum topology, similarity measure, or learning mode. In medicine, biology, and medical bioinformatics, more and more data arise from clinical measurements such as EEG or fMRI studies for monitoring brain activity, mass spectrometry data for the detection of proteins, peptides and composites, or microarray profiles for the analysis of gene expressions. Typically, data are high-dimensional, noisy, and very hard to inspect using classic (e.g., symbolic or linear) methods. At the same time, new technologies ranging from the possibility of a very high resolution of spectra to high-throughput screening for microarray data are rapidly developing and carry the promise of an efficient, cheap, and automatic gathering of tons of high-quality data with large information potential. Thus, there is a need for appropriate machine learning methods which help to automatically extract and interpret the relevant parts of this information and which, eventually, help to enable understanding of biological systems, reliable diagnosis of faults, and therapy of diseases such as cancer based on this information. Moreover, these application scenarios pose fundamental and qualitatively new challenges to the learning systems because of the specifics of the data and learning tasks. Since these characteristics are particularly pronounced within the medical domain, but not limited to it and of principled interest, this research topic opens the way toward important new directions of algorithmic design and accompanying theory.

Similarity-based learning models are ideally suited for an application in medical and biological domains or application areas which incorporate related settings because of several crucial aspects including the possibility of human insight into their behavior, a large flexibility of the models, and an excellent generalization ability also for high-dimensional data. Several successful applications ranging from the visualization of microarray profiles up to cancer prediction demonstrate this fact. However, some effort will still be needed to develop reliable, efficient, and user-friendly machine-learning tools and the corresponding theoretical background to really suit the specific needs in medical applications: on the one hand, this application area poses specific requests on interpretability and accuracy of the models and their usability for people without knowledge in machine learning (e.g., physicians), on the other hand, typical problems and data structures in the medical domain provide additional information (e.g., prior knowledge about proteins, the specific form of a spectrum, or the relevance of certain data features), which could be included into the learning methods to shape the output according to the specific situation.

In Spring 2007, 33 scientist from 10 different countries gathered together in Dagstuhl Castle in the south of Germany to discuss important new scientific developments and challenges in the frame of unsupervised clustering, in particular in the context of applications in life science. According to the interdisciplinary topic, researchers came from different disciplines including pattern recognition and machine learning, theoretical computer science, statistical physics, biology, and medicine. Within this highly stimulating and interdisciplinary environment, several topics at the frontiers of research were discussed concerning a theoretical investigation and foundation of prototype-based learning algorithms, the development and extension of models to new challenging directions such as general data structures, and the application for the domain of medicine and biology. While often tackled separately, these three stages were discussed together to judge their mutual influence and to further an integration of different aspects to achieve optimum models, algorithmic design, and theoretical background. This book has emerged as a summary and result of the interdisciplinary meeting, and it gathers together overview articles about recent developments, trends, and applications of similarity-based learning toward biomedical applications and beyond. In three chapters, the three fundamental aspects of a theoretical background, the representation of data and their connection to algorithms, and their particular challenging applications are considered.

More precisely, the first chapter deals with the objectives of clustering and the dynamics of similarity-based learning. Clustering is an inherently ill-posed problem, and optimal solutions depend on the concrete task at hand. Therefore, a variety of different methods exist that are derived on the basis of general mathematical principles or even only of intuitive heuristics. Correspondingly, an exact mathematical investigation of the methods is crucial to judge the performance and reliability of the methods, but at the same time it is often very difficult. Further, even if a concrete objective of clustering is given, the learning dynamics can be highly nontrivial and algorithmic optimization might be necessary to arrive at good solutions of the problem. The first article by Biehl, Caticha, and Riegler gives an overview of the possibility to exactly investigate the learning dynamics of similarity-based learning rules by means of the so-called theory of online learning, which is heavily based on statistical physics. A specific learning rule, neural gas, constitutes the focus of the second contribution by Villmann, Hammer, and Biehl, and possibilities to extend the dynamics to general metrics as well as the connection of topographic coordination to deterministic annealing are discussed. Finally, Fyfe and Barbakh propose an alternative algorithmic solution to get around local minima which consitute a major problem for popular clustering algorithms such as k-means and self-organizing maps. As an alternative, they adapt reinforcement learning algorithms and dynamic programming toward these tasks.

The second chapter addresses another fundamental problem of clustering: the issue of a correct representation of data. In supervised algorithms, data are observed through the interface of a metric, and this metric has to be chosen appropriately to allow efficient and reliable training. Verleysen, Rossi, and Damien consider the important issue of selecting relevant features if the euclidean metric is used since, otherwise, a failure due to the accumulation of noise can be observed particularly for high-dimensional data sets as often given in biomedical applications. Instead of the euclidean meric, Pearson correlation can be used for non-normalized data with benefitial effects if, for example, the overall shape is more important than the size of the values. This topic is discussed in applications from bioinformatics by Strickert, Schleif, Villmann, and Seiffert. Often, in biological domains, no explicit (differentiable) metric is available at all, rather, data are characterized by pairwise dissimilarities only. Hammer, Hasenfuss, and Rossi discuss extensions of clustering and topographic mapping toward general dissimilarity data by means of the generalized median. Thereby, particular emphasis is laid on the question of how an algorithmic speed-up becomes possible to cope with realistic data sets. Finally, Giannotis and Tino consider fairly general data structures, sequential and graph-structured data, which can be embedded into similarity-based algorithms by means of general probabilistic models that describe the data.

Similarity-based methods find widespread applications in diverse application domains, including in particular biomedical problems, but also geophysics or technical domains. In all areas, however, a number of challenges still have to be faced and a straightforward application of standard algorithms using the euclidean metric is only rarely possible. This observation is demonstrated in the third chapter, which presents discussions about different challenging real-life applications of similarity-based methods. Merenyi, Tademir, and Zhang explain particular problems that occur when very complex data manifolds have to be addressed as occurs, for example, in remote sensing and geoscience applications. Using large-scale data, they demonstrate the problems of extracting true clusters from the data and of evaluating the trustworthiness of the results. Sanchez and Petkov investigate the problem of defining a relevant problem-dependent metric in a biomedical application in image analysis. Ineterstingly, a correct (nontrivial) choice of the metric allows use of comparably simple subsequent processing and it can constitute the core problem in some settings. The final contribution by Rosen-Zvi, Aharni, and Selbig presents state-of-the-art research when addressing challenging large-scale problems in bioinformatics, in this case the prediction of HIV-1 drug resistance. In an impressive way, the question of how to deal with high dimensionality and missing values, among others, is discussed.

Altogether, these presentations give a good overview about important research results in similarity-based learning concerning theory, algorithmic design, and applications, whereby the character of the overview articles with references to correlated research articles makes the contributions particularly suited for a first reading concerning these topics. Since many challenging problems still lie ahead and research will evolve methods beyond this state of the art, there will certainly be a replication of seminars on this topics.

October 2008

Michael Biehl
Barbara Hammer
Michel Verleysen
Thomas Villmann

Table of Contents

Table of Contents

Statistical Mechanics of On-line Learning

Michael Biehl[1], Nestor Caticha[2], and Peter Riegler[3]

[1] University of Groningen, Institute of Mathematics and Computing Science,
P.O. Box 407, 9700 AK Groningen, The Netherlands
[2] Instituto de Fisica, Universidade de São Paulo,
CP66318, CEP 05315-970, São Paulo, SP Brazil
[3] Fachhochschule Braunschweig/Wolfenbüttel, Fachbereich Informatik,
Salzdahlumer Str. 46/48, 38302 Wolfenbüttel, Germany

Abstract. We introduce and discuss the application of statistical physics concepts in the context of on-line machine learning processes. The consideration of typical properties of very large systems allows to perfom averages over the randomness contained in the sequence of training data. It yields an exact mathematical description of the training dynamics in model scenarios. We present the basic concepts and results of the approach in terms of several examples, including the learning of linear separable rules, the training of multilayer neural networks, and Learning Vector Quantization.

1 Introduction

The perception that Statistical Mechanics is an inference theory has opened the possibility of applying its methods to several areas outside the traditional realms of Physics. This explains the interest of part of the Statistical Mechanics community during the last decades in machine learning and optimization and the application of several techniques of Statistical Mechanics of disordered systems in areas of Computer Science.

As an inference theory Statistical Mechanics is a Bayesian theory. Bayesian methods typically demand extended computational resources which probably explains why methods that were used by Laplace, were almost forgotten in the following century. The current availability of ubiquitous and now seemingly powerful computational resources has promoted the diffusion of these methods in statistics. For example, posterior averages can now be rapidly estimated by using efficient Markov Chain Monte Carlo methods, which started to be developed to attack problems in Statistical Mechanics in the middle of last century. Of course the drive to develop such techniques (starting with [1]) is due to the total impossibility of introducing classical statistics methods to study thermodynamic problems. Several other techniques from Statistical Mechanics have also found their way into statistics. In this paper we review some applications of Statistical Mechanics to artificial machine learning.

Learning is the change induced by the arrival of information. We are interested in learning from examples. Different scenarios arise if examples are considered in

M. Biehl et al.: (Eds.): Similarity-Based Clustering, LNAI 5400, pp. 1–22, 2009.

batches or just one at a time. This last case is the on-line learning scenario and the aim of this contribution is to present the characterization of on-line learning using methods of Statistical Mechanics. We will consider in section 2 simple linearly separable classification problems, just to introduce the idea. The dynamics of learning is described by a set of stochastic difference equations which show the evolution, as information arrives, of quantities that characterize the problem and in Statistical Mechanics are called order parameters. Looking at the limit of large dimensions and scaling the number of examples in a correct manner, the difference equations simplify into deterministic ordinary differential equations. Numerical solutions of the ODE give then, for the specific model of the available information such as the distribution of examples, the learning curves.

While this large dimension limit may seem artificial, it must be stressed that it is most sensible in the context of thermostatistics, where Statistical Mechanics is applied to obtain thermodynamic properties of physical systems. There the dimension, which is the number of degrees of freedom, is of the order of Avogadro's number ($\approx 10^{23}$). Simulations however have to be carried for much smaller systems, and this prompted the study of finite size corrections of expectations of experimentally relevant quantities, which depend on several factors, but typically go to zero algebraically with the dimension. If the interest lies in intensive quantities such as temperature, pressure or chemical potential then corrections are negligible. If one is interested on extensive quantities, such as energy, entropy or magnetization, one has to deal with their densities, such as energy per degree of freedom. Again the errors due to the infinite size limit are negligible. In this limit, central limit theorems apply, resulting in deterministic predictions and the theory is in the realm of thermodynamics. Thus this limit is known as the thermodynamic limit (TL). For inference problems we can use the type of theory to control finite size effects in the reverse order. We can calculate for the easier deterministic infinite limit and control the error made by taking the limit. We mention this and give references, but will not deal with this problem except by noticing that it is theoretically understood and that simulations of stylized models, necessarily done in finite dimensions, agree well with the theoretical results in the TL. The reader might consider that the thermodynamic limit is equivalent to the limit of infinite sequences in Shannon's channel theorem.

Statistical Mechanics (see e.g. [2,3]) had its origins in the second half of the XIX century, in an attempt, mainly due to Maxwell, Boltzmann and Gibbs to deduce from the microscopic dynamical properties of atoms and molecules, the macroscopic thermodynamic laws. Its success in building a unified theoretical framework which applied to large quantities of experimental setups was one of the great successes of Physics in the XIX century. A measure of its success can be seen from its role in the discovery of Quantum Mechanics. Late in the XIX century, when its application to a problem where the microscopic laws involved electromagnetism, i.e the problem of Black Body radiation, showed irreconcilable results with experiment, Max Planck showed that the problem laid with the classical laws of electromagnetism and not with Statistical Mechanics. This started the revolution that led to Quantum Mechanics.

This work is organized as follows. We introduce the method in section 2 the main ideas in a simple model. Section 3 will look into the extension of on-line methods to the richer case of two-layered networks which include universal approximators. In section 4 we present the latest advances in the area which deal with characterizing theoretically clustering methods such as Learning Vector Quantization (LVQ).

This paper is far from giving a complete overview of this successful approach to machine learning. Our intention is to illustrate the basic concepts in terms of selected examples, mainly from our own work. Also references are by no means complete and serve merely as a starting point for the interested reader.

2   On-line Learning in Classifiers: Linearly Separable Case

We consider a supervised classification problem, where vectors $\boldsymbol{\xi} \in \mathbf{R}^N$ have to be classified into one of two classes which we label by $+1$ or -1 respectively. These vectors are drawn independently from a probability distribution $P_o(\boldsymbol{\xi})$. The available information is in the form of example pairs of vector-label: $(\boldsymbol{\xi}_\nu, \sigma_\nu)$, $\nu = 1, 2...\mu$. The scenario of on-line learning is defined by the fact that we take into account one pair at a time, which permits to identify ν and μ as time indexes. We also restrict our attention to the simple case where examples are used once to induce some change in our machine and then are discarded. While this seems quite inefficient since recycling examples to extract more information can indeed be useful, it permits to develop a simple theory due to the assumption of independence of the examples. The recycling of examples can also be treated ([4]) but it needs a repertoire of techniques that is beyond the scope of this review. For many simple cases this will be seen to be quite efficient.

As a measure of the efficiency of the learning algorithm we will concentrate on the generalization error, which is the probability of making a classification error on a new, statistically independent example $\boldsymbol{\xi}_{\mu+1}$. If any generalization is at all possible, of course there must be an underlying rule to generate the example labels, which is either deterministic

$$\sigma^B = f_B(\boldsymbol{\xi}) \tag{1}$$

or described by the conditional probability $P(\sigma^B | f_B(\boldsymbol{\xi}))$ depending on a transfer function f_B parameterized by a set of K unknown parameters B. At this point we take B to be fixed in time, a constraint that can be relaxed and still be studied within the theory, see [5,6,7,8].

Learning is the compression of information from the set of μ example pairs into a set of M weights $\boldsymbol{J}_\mu \in \mathbf{R}^M$ and our machine classifies according to

$$\sigma^J = g_J(\boldsymbol{\xi}) \tag{2}$$

The generalization error is

$$e_G(\mu) = \int dP_o(\boldsymbol{\xi}) \int \prod_{\nu=1}^{\mu} dP_o(\boldsymbol{\xi}_\nu) \sum_{\sigma_\nu = \pm 1} P(\sigma_\nu^B | f_B(\boldsymbol{\xi}_\nu)) \Theta(-\sigma^{J_\mu} \sigma^B(\boldsymbol{\xi}))$$

$$= \langle \Theta(-\sigma^{J_\mu}(\boldsymbol{\xi})\sigma^B(\boldsymbol{\xi})) \rangle_{\{\sigma_\nu, \boldsymbol{\xi}_\nu\}_{\nu=1}^\mu, \sigma, \boldsymbol{\xi}}, \tag{3}$$

where the step function $\Theta(x) = 1$ for $x > 0$ and zero otherwise. As this stands it is impossible to obtain results other than of a general nature. To obtain sharp results we have to specify a model.

The transfer functions f_B and g_J specify the architectures of the rule and of the classifier, while $P(\sigma_\mu^B | f_B(\boldsymbol{\xi}_\mu))$ models possible noise in the specification of the supervision label. The simplest case that can be studied is where both f_B and g_J are linearly separable classifiers of the same dimension: $K = M = N$,

$$\sigma^J = \text{sign}(\boldsymbol{J}.\boldsymbol{\xi}), \qquad \sigma^B = \text{sign}(\boldsymbol{B}.\boldsymbol{\xi}) \tag{4}$$

As simple and artificial as it may be, the study of this special case serves several purposes and is a stepping stone into more realistic scenarios.

An interesting feature of Statistical Mechanics lies in that it points out what are the relevant order parameters in a problem. In physics, this turns out to be information about what are the objects of experimental interest.

Without any loss we can take all vectors $\boldsymbol{\xi}$ and $\boldsymbol{B}$ to be normalized as $\boldsymbol{\xi}\cdot\boldsymbol{\xi} = N$ and $\boldsymbol{B}\cdot\boldsymbol{B} = 1$. For $\boldsymbol{J}$ however, which is a dynamical quantity that evolves under the learning algorithm still to be specified we let it free and call $\boldsymbol{J} \cdot \boldsymbol{J} = Q$. Define the fields

$$h = \boldsymbol{J}\cdot\boldsymbol{\xi}, \qquad b = \boldsymbol{B}\cdot\boldsymbol{\xi} \tag{5}$$

To advance further we choose a model for the distribution of examples $P_o(\boldsymbol{\xi})$ and the natural starting point is to choose a uniform distribution over the N-dimensional sphere. Different choices to model specific situation are of course possible. Under these assumptions, since the scalar products of (4) are sums of random variables, for large N, h and b are correlated Gaussian variables, completely characterized by

$$\langle h \rangle = \langle b \rangle = 0,$$
$$\langle h^2 \rangle = Q, \quad \langle b^2 \rangle = 1,$$
$$\langle hb \rangle = \boldsymbol{J} \cdot \boldsymbol{B} = R. \tag{6}$$

It is useful to introduce the overlap $\rho = R/\sqrt{Q}$ between the rule and machine parameter vectors. The joint distribution is given by

$$P(h, b) = \frac{1}{2\pi\sqrt{(1-\rho^2)}} e^{-\frac{1}{2(1-\rho^2)}(h^2 - 2\rho hb + b^2)}. \tag{7}$$

The correlation is the overlap ρ, which is related to the angle ϕ between $\boldsymbol{J}$ and $\boldsymbol{B}$: $\phi = \cos^{-1}\rho$, it follows that $|\rho| \leq 1$. It is geometrically intuitive and also easy to prove that the probability of making an error on an independent example, the generalization error, is $\phi/2\pi$:

$$e_G = \frac{1}{2\pi} \cos^{-1}\rho \tag{8}$$

The strategy now is to introduce a learning algorithm, i.e to define the change that the inclusion of a new example causes in J, calculate the change in the overlap ρ and then obtain the learning curve for the generalization error. We will consider learning algorithms of the form

$$J_{\mu+1} = J_\mu + \frac{F}{N}\boldsymbol{\xi}_{\mu+1}, \tag{9}$$

where F, called the modulation function of vector $\boldsymbol{\xi}_{\mu+1}$, should depend on the supervised information, the label $\sigma^B_{\mu+1}$. It may very well depend on some additional information carried by hyperparameters or on $\boldsymbol{\xi}$ itself. It is F that defines the learning algorithm. We consider the case where F is a scalar function, but it could differ for different components of $\boldsymbol{\xi}$. Projecting (9) into $\boldsymbol{B}$ and into $\boldsymbol{J}$ we obtain respectively

$$R_{\mu+1} = R_\mu + \frac{F}{N}b_\mu \tag{10}$$

$$Q_{\mu+1} = Q_\mu + 2\frac{F}{N}h_\mu + \frac{F^2}{N}. \tag{11}$$

which describe the learning dynamics. We can also write an equivalent equation for the overlap ρ which is valid for large N and $\boldsymbol{\xi}$ on the hypersphere and

$$\rho_{\mu+1} = \frac{J_{\mu+1}.B}{\sqrt{Q_{\mu+1}}} =$$

$$= \rho_\mu \left(1 - \frac{1}{N}\frac{F}{\sqrt{Q_\mu}}h_{\mu+1} - \frac{1}{2N}\left(\frac{F}{\sqrt{Q_\mu}}\right)^2\right) + \frac{1}{N}\frac{F}{\sqrt{Q_\mu}}b_{\mu+1} \tag{12}$$

$$\Delta\rho_{\mu+1} = \frac{1}{N}\frac{F}{\sqrt{Q_\mu}}(b_{\mu+1} - \rho_\mu h_{\mu+1}) - \frac{1}{2N}\frac{\rho_\mu F^2}{Q_\mu} \tag{13}$$

Since at each time step μ a random vector is drawn from the distribution P_o equations (12) and (13) are stochastic difference equations. We now take the thermodynamic limit $N \to \infty$ and average over the test example $\boldsymbol{\xi}$. Note that each example induces a change of the order parameters of order $1/N$. Hence, one expects the need of order N many examples to create a change of order 1. This prompts the introduction of $\alpha = \lim_{N\to\infty}\mu/N$ which by measuring the number of examples measures time. The behavior of ρ and $\frac{\Delta\rho}{\Delta\alpha}$ are very different in the limit. It can be shown (see [9]) that order parameters such as ρ, R, Q self-average. Their fluctuations tend to zero in this limit, see Fig. 3 for an example. On the other hand $\frac{\Delta\rho}{\Delta\alpha}$ has fluctuations of order one and we look at its average over the test vector:

$$\frac{d\rho}{d\alpha} = \left\langle\frac{\Delta\rho}{\Delta\alpha}\right\rangle_{h,b,\xi}. \tag{14}$$

the pairs $(Q, \Delta Q/\Delta\alpha)$ and $(R, \Delta R/\Delta\alpha)$ behave in a similar way. We average over the fields h, b and over the labels σ, using $P(\sigma|b)$ .This leads to the coupled system of ordinary differential equations, which for a particular form of F were introduced in [10].

$$\frac{dR}{d\alpha} = \sum_\sigma \int dh db P(h,b) P(\sigma|b) [Fb] = \langle Fb \rangle \tag{15}$$

$$\frac{dQ}{d\alpha} = \sum_\sigma \int dh db P(h,b) P(\sigma|b) [2Fh + F^2] = \langle 2Fh + F^2 \rangle, \tag{16}$$

where the angular brackets stand for the average over the fields and label. Since the generalization error is directly related to ρ it will be useful to look at the equivalent set of equations for ρ and the length of the weight vector $\sqrt{Q}$:

$$\frac{d\rho}{d\alpha} = \sum_\sigma \int dh db P(h,b) P(\sigma|b) \left[\frac{F}{\sqrt{Q}} (b - \rho h) - \frac{\rho F^2}{2Q} \right] \tag{17}$$

$$\frac{d\sqrt{Q}}{d\alpha} = \sum_\sigma \int dh db P(h,b) P(\sigma|b) \left[Fh + \frac{1}{2} \frac{F^2}{\sqrt{Q}} \right] \tag{18}$$

We took the average with respect to the two fields h and b as if they stood on symmetrical grounds, but they don't. It is reasonable to assume knowledge of h and σ but not of b. Making this explicit

$$\frac{d\rho}{d\alpha} = \sum_\sigma \int dh P(h) P(\sigma) \left[\frac{F}{\sqrt{Q}} \langle b - \rho h \rangle_{b|\sigma h} - \frac{\rho F^2}{2Q} \right] \tag{19}$$

$$\frac{d\sqrt{Q}}{d\alpha} = \sum_\sigma \int dh P(h) P(\sigma) \left[Fh + \frac{1}{2} \frac{F^2}{\sqrt{Q}} \right] \tag{20}$$

call

$$F^* = \frac{\sqrt{Q}}{\rho} \langle b - \rho h \rangle_{b|\sigma h} \tag{21}$$

where the average is over unavailable quantities. Equation 19 can then be written as

$$\frac{d\rho}{d\alpha} = \frac{\rho}{Q} \sum_\sigma \int dh P(h) P(\sigma) \left[F F^* - \frac{1}{2} F^2 \right] \tag{22}$$

This notation makes it natural to ask for an interpretation of the meaning of F^*. The differential equations above describe the dynamics for *any* modulation function. We can ask ([11]) if there is a modulation function optimal in the sense of maximizing the information gain per example, which can be stated as a variational problem

$$\frac{\delta}{\delta F} \left(\frac{d\rho}{d\alpha} \right) = 0 \tag{23}$$

This problem can be solved for a few architectures, including some networks with hidden layers [12,16], although the optimization becomes much more difficult.

Within the class of algorithms we have considered, the performance bound is given by the modulation function F^* given by Eq. (21).

The optimal resulting algorithm has several features that are useful in considering practical algorithms, such as the optimal annealing schedule of the learning rate and, in the presence of noise, adaptive cutoffs for surprising examples.

As an example we discuss learning of a linearly separable rule (4) in some detail. There, an example will be misclassified if $\theta(-\sigma^J \sigma^B) > 0$ or equivalently $h\sigma^B < 0$. We will refer to the latter quantity as aligned field. It basically measures the correctness ($h\sigma^B > 0$) or wrongness ($h\sigma^B < 0$) of the current classification.

Fig. 1 depicts the optimal modulation function for inferring a linearly separable rule from linearly separable data. Most interesting is the dependence on $\rho = \cos(\pi\epsilon_g)$: For $\rho = 0$ ($\epsilon_g = 1/2$) the modulation function does not take into account the correctness of the current examples. The modulation function is constant, which corresponds to Hebbian learning. As ρ increases, however, for already correctly classified examples the magnitude of the modulations function decreases with increasing aligned field. For misclassified examples, however, the update becomes the larger the smaller the aligned field is. In the limit $\rho \to 1$ ($\epsilon_g \to \infty$) the optimal modulation function approaches the Adatron algorithm ([13]) where only misclassified examples trigger an update which is proportional to the aligned field. In addition to that, for the optimal modulation function $\rho(\alpha) = \sqrt{Q(\alpha)}$, i.e. the order parameter Q can be used to estimate ρ.

Now imagine, that the label of linearly separable data is noisy, i.e. it is changed with a certain probability. In this case it would be very dangerous to follow an Adatron-like algorithm and perform an large update if $h\sigma^B < 0$, since the seeming misclassification might be do to a corrupted label. The optimal modulation function for that case perceives this danger and introduces a sort of cutoff w.r.t. the aligned field. Hence, no considerable update is performed if the aligned field becomes too large in magnitude. Fig. 1 shows the general behavior. [14] gives further details and an extension to other scenarios of noisy but otherwise linearly separable data sets.

These results for optimal modulations functions are, in general, better understood from a Bayesian point of view ([17,18,19]). Suppose the knowledge about the weight vector is coded in a Gaussian probability density. As a new example arrives, this probability density is used as a prior distribution. The likelihood is built out of the knowledge of the network architecture, of the noise process that may be corrupting the label and of the example vector and its label. The new posterior is not, in general Gaussian and a new Gaussian is chosen, to be the prior for the next learning step, in such a way as to minimize the information loss. This is done by the maximum entropy method or equivalently minimizing the Kullback-Leibler divergence. It follows that the learning of one example induces a mapping from a Gaussian to another Gaussian, which can be described by update equations of the Gaussian's mean and covariance. These equations define a learning algorithm together with a tensor like adaptive schedule for the learning rate. This algorithms are closely related to the variational algorithm

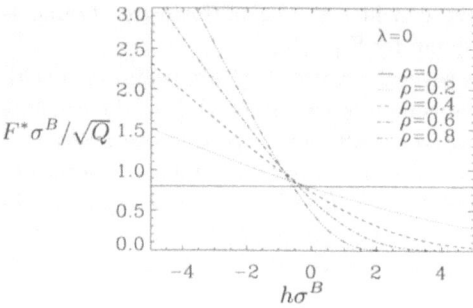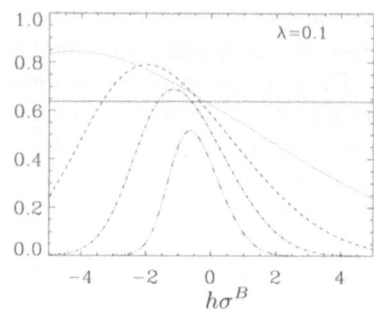

Fig. 1. Left: Optimal modulation function F^* for learning a linearly separable classifi-cation of type (4) for various values of the order parameter ρ. Right: In addition to (4) the labels σ^B are subject to noise, i.e. are flipped with probability λ. The aligned field $h\sigma^B$ is a measure of the agreement of the classification of the current example with the given label. In both cases, all examples are weighted equally for $\rho = 0$ irrespective of the value of the aligned field $h\sigma^B$. This corresponds to Hebbian learning. In the noiseless case ($\lambda = 0$) examples with a negative aligned field receive more weight as ρ increases while those with positive aligned field gain less weight. For $\lambda > 0$ also examples with large negative aligned fields are not taken into account for updating. These examples are most likely affected by the noise and therefore are deceptively misclassified. The optimal weight function possesses a cutoff value at negative values of $h\sigma^B$. This cutoff decreases in absolute value with increasing λ and ρ.

defined above. The Bayesian method has the advantage of being more general. It can also be readily applied to other cases where the Gaussian prior is not adequate [19].

   In the following sections we follow the developments of these techniques in other directions. We first look into networks with hidden layers, since the expe-rience gained in studying on-line learning in these more complex architectures will be important to better understand the application of on-line learning to the problem of clustering.

3   On-line Learning in Two-Layered Networks

Classifiers such as (4) are the most fundamental learning networks. In terms of network architecture the next generalization are networks with one layer of hidden units and a fixed hidden-to-output relation. An important example is the so-called committee machine, where the overall output is determined by a majority vote of several classifiers of type (4). For such networks the variational approach of the previous section can be readily applied [12,16].

   General two-layered networks, however, have variable hidden-to-output weights and are "soft classifiers", i.e. have continuous transfer functions. They consist of a layer of N inputs, a layer of K hidden units, and a single output. The particular input-output relation is given by

$$\sigma(\boldsymbol{\xi}) = \sum_{i=1}^{K} w_i g(\boldsymbol{J}_i \cdot \boldsymbol{\xi}). \tag{24}$$

Here, $\boldsymbol{J}_i$ denotes the N-dimensional weight vector of the i-th input branch and w_i the weight connecting the i-th hidden unit with the output. For a (soft) committee machine, $w_i \equiv 1$ for all branches i. Often, the number K of hidden weight vectors $\boldsymbol{J}_i$ is chosen as $K \ll N$. In fact, most analyses specialize on $K = \mathcal{O}(1)$. This restriction will also be pursued here. Note that the overall output is linear in w_i, in contrast to the outputs of the hidden layer which in general depend nonlinearly on the weights $\boldsymbol{J}_i$ via the transfer function g.

The class of networks (24) is of importance for several reasons: Firstly, they can realize more complex classification schemes. Secondly, they are commonly used in practical applications. Finally, two-layered networks with a linear output are known to be universal approximators [20].

Analogously to section 2 the network (24) is trained by a sequence of uncorrelated examples $\{(\boldsymbol{\xi}^\nu, \tau^\nu)\}$ which are provided by an unknown rule $\tau(\boldsymbol{\xi})$ by the environment. As above, the example input vectors are denoted by $\boldsymbol{\xi}^\mu$, while here τ^μ is the corresponding correct rule output.

In a commonly studied scenario the rule is provided by a network of the same architecture with hidden layer weights $\boldsymbol{B}_i$, hidden-to-output weights v_i, and an in general different number of hidden units M:

$$\tau(\boldsymbol{\xi}) = \sum_{k=1}^{M} v_k g(\boldsymbol{B}_k \cdot \boldsymbol{\xi}). \tag{25}$$

In principle, the network (24) can implement such a function if $K \geq M$.

As in (9) the change of a weight is usually taken proportional to the input of the corresponding layer in the network, $i.e.$

$$\boldsymbol{J}_i^{\mu+1} = \boldsymbol{J}_i^\mu + \frac{1}{N} F_i \boldsymbol{\xi}^{\mu+1} \tag{26}$$

$$w_i^{\mu+1} = w_i^\mu + \frac{1}{N} F_w g(\boldsymbol{J}_i \cdot \boldsymbol{\xi}^{\mu+1}). \tag{27}$$

Again, the modulation functions F_i, F_w will in general depend on the recently provided information $(\boldsymbol{\xi}^\mu, \tau^\mu)$ and the current weights.

Note, however, that there is an asymmetry between the updates of $\boldsymbol{J}_i$ and w_i. The change of the former is $\mathcal{O}(1/N)$ due to $|\boldsymbol{\xi}^2| = \mathcal{O}(N)$. As $\sum_{i=1}^{K} g^2(\boldsymbol{J}_i \cdot \boldsymbol{\xi}) = \mathcal{O}(K)$ a change of the latter according to

$$w_i^{\mu+1} = w_i^\mu + \frac{1}{K} F_w g(\boldsymbol{J}_i \cdot \boldsymbol{\xi}^{\mu+1}). \tag{28}$$

seems to be more reasonable. For reasons that will become clear below we will prefer a scaling with $1/N$ as in (27) over a scaling with $1/K$, at least for the time being.

Also note from (24, 25) that the stochastic dynamics of $\boldsymbol{J}_i$ and w_i only depends on the fields $h_i = \boldsymbol{J}_i \cdot \boldsymbol{\xi}$, $b_k = \boldsymbol{B}_k \cdot \boldsymbol{\xi}$ which can be viewed as a generalization of Eq. (5). As in section 2, for large N these become Gaussian variables. Here, they have zero means and correlations

$$\langle h_i h_j \rangle = \boldsymbol{J_i} \cdot \boldsymbol{J_j} =: Q_{ij} \ , \quad \langle b_k b_l \rangle = \boldsymbol{B_k} \cdot \boldsymbol{B_l} =: T_{kl} \ , \quad \langle h_i b_k \rangle = \boldsymbol{J_i} \cdot \boldsymbol{B_k} =: R_{ik}, \quad (29)$$

where $i, j = 1 \ldots K$ and $k, l = 1 \ldots M$.

Introducing $\alpha = \mu/N$ as above, the discrete dynamics (26, 27) can be replaced by a set of coupled differential equations for R_{ik}, Q_{ij}, and w_i in the limit of large N: Projecting (26) into $\boldsymbol{B_k}$ and $\boldsymbol{J_j}$, respectively, and averaging over the randomness of $\boldsymbol{\xi}$ leads to

$$\frac{dR_{ik}}{d\alpha} = \langle F_i b_k \rangle \tag{30}$$

$$\frac{dQ_{ij}}{d\alpha} = \langle F_i h_j + F_j h_i + F_i F_j \rangle, \tag{31}$$

where the average is now with respect to the fields $\{h_i\}$ and $\{b_k\}$. Hence, the microscopic stochastic dynamics of $\mathcal{O}(K \cdot N)$ many weights $\boldsymbol{J_i}$ is replaced by the macroscopic dynamics of $\mathcal{O}(K^2)$ many order parameters R_{ik} and Q_{ij}, respectively. Again, these order parameters are self-averaging, *i.e.* their fluctuations vanish as $N \to \infty$. Fig. 3 exemplifies this for a specific dynamics.

The situation is somewhat different for the hidden-to-output weights w_i. In the transition from microscopic, stochastic dynamics to macroscopic, averaged dynamics the hidden-layer weights $\boldsymbol{J_i}$ are compressed to order parameters which are scalar products, cf. (29). The hidden-to-output weights, however, are not compressed into new parameters of the form of scalar products. (Scalar products of the type $\sum_i w_i v_i$ do not even exist for $K \neq M$.) Scaling the update of w_i by $1/N$ as in (27) allows to replace $1/N$ by the differential $d\alpha$ as $N \to \infty$. Hence, the averaged dynamics of the hidden-to-output weight reads

$$\frac{dw_i}{d\alpha} = \langle F_w g(h_i) \rangle. \tag{32}$$

Note that the r.h.s. of these differential equations depend on R_{ik}, Q_{ij} via (29) and, hence, are coupled to the differential equations (30, 31) of these order parameters as well. So in total, the macroscopic description of learning dynamics consists of the coupled set (30, 31, 32).

It might be surprising that the hidden-to-output weights w_i by themselves are appropriate for a macroscopic description while the hidden weights $\boldsymbol{J_i}$ are not. The reason for this is twofold. First, the number K of w_i had been taken to be $\mathcal{O}(1)$, *i.e.* it does not scale with the dimension of inputs N. Therefore, there is no need to compress a large number of microscopic variables into a small number of macroscopic order parameters as for the $\boldsymbol{J_i}$. Second, the change in w_i had been chosen to scale with $1/N$. For this choice one can show that like R_{ik} and Q_{ij} the weights w_i are self-averaging.

For a given rule $\tau(\boldsymbol{\xi})$ to be learned, the generalization error is

$$\epsilon_g(\{\boldsymbol{J_i}, w_i\}) = \langle \epsilon(\{\boldsymbol{J_i}, w_i\}) \rangle_{\boldsymbol{\xi}}, \tag{33}$$

where $\epsilon(\{\boldsymbol{J_i}, w_i\}) = \frac{1}{2}(\sigma - \tau)^2$ is the instantaneous error. As the outputs σ and τ depend on $\boldsymbol{\xi}$ only via the fields $\{h_i\}$ and $\{b_k\}$, respectively, the average over $\boldsymbol{\xi}$

can be replaced by an average over these fields. Hence, the generalization error only depends on order parameters R_{ik}, Q_{ij}, w_i as well as on T_{kl} and v_k.

In contrast to section 2 it is a difficult task to derive optimal algorithms by a variational approach, since the generalization error (33) is a function of several order parameters. Therefore on-line dynamics in two-layer networks has mostly been studied in the setting of heuristic algorithms, in particular for backpropagation. There, the update of the weights is taken proportional to the gradient of the instantaneous error $\epsilon = \epsilon(\{J_i, w_i\}) = \frac{1}{2}(\sigma - \tau)^2$ with respect to the weights:

$$J_i^{\mu+1} = J_i^{\mu} - \frac{\eta_J}{N} \nabla_{J_i} \epsilon \tag{34}$$

$$w_i^{\mu+1} = w_i^{\mu} - \frac{\eta_w}{N} \frac{\partial \epsilon}{\partial w_i} \tag{35}$$

The parameters η_J and η_w denote learning rates which scale the size of the update along the gradient of the instantaneous error. In terms of (26, 27) back-propagation corresponds to the special choices $F_J = -\eta_J(\sigma - \tau)w_i g'(h_i)$ and $F_w = -\eta_w(\sigma - \tau)$. A common choice for the transfer function is $g(x) = \mathrm{erf}(x/\sqrt{2})$. With this specific choice, the averaging in the equations of motion (30, 31, 32) can be performed analytically for general K and M [22,23,24].

Independent of the particular choice of learning algorithms a general problem in two-layered networks is caused by the inherent permutation symmetry: The i-th input branch of the adaptive network (24) does not necessarily specialize on the i-th branch in the network (25). Without loss of generality, however, one can relabel the dynamical variables such as if this were indeed the case. Nevertheless, this permutation symmetry will turn out to be a dominant feature because it leads to a deceleration of learning due to plateau states in the dynamics of the generalization error. Fig. 2 gives an example.

These plateau states correspond to configurations which are very close to certain fixed points of the set of differential equations (30, 31, 32) for the order

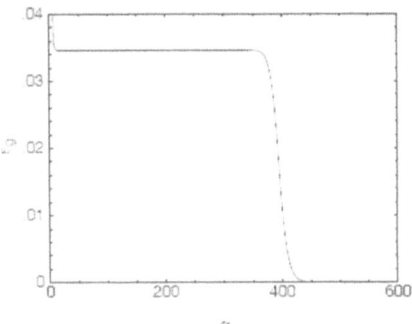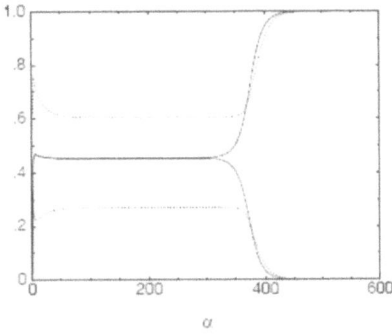

Fig. 2. Time evolution for on-line backpropagation in the $K = M = 2$ learning scenario (24,25,34) with $T_{nm} = \delta_{nm}$ and fixed $w_i = v_i$. Left: generalization error $\epsilon_g(\alpha)$. Right: order parameters R_{in} (full curves) and Q_{ik} (dotted curves). The plateaus in both graphs are due to the internal permutation symmetry of(25) w.r.t. the summation index.

parameters. In the simplest case the vectors J_i are almost identical during the plateau phase. They have – apart from small deviations – the same scalar product with each vector B_k. These fixed points are repulsive, so small fluctuations will cause a specialization of each J_i towards a distinct B_k which then leads to minimum generalization error. If there are several such repulsive fixed points there can even be cascades of plateaus. The lengths of the plateaus can be shown to depend crucially on the choice of initial conditions as well as on the dimension N of the inputs [25].

For backpropagation, the differential equations (30, 31, 32) can easily be used to determine those learning rates η_J and η_w which lead to the fastest decrease of the generalization error. This is most interesting for the learning rate η_w of the hidden-to-output weights as it turns out that the decrease of $\epsilon_g(\alpha)$ is largest as $\eta_w \to \infty$.

Obviously, this divergence of η_w indicates that one should have chosen a different scaling for the change of the weights w_i, namely a scaling with $1/K$ as in (28) as opposed to (27). For such a scaling, the weights w_i will not be self-averaging anymore, however, see Fig. 3. Hence, equations (30, 31, 32) fail to provide a macroscopic description of the learning dynamics in this case. This does by no means signify that they are inapplicable, however. The optimal choice $\eta_w \to \infty$ simply indicates that the dynamics of the hidden-to-output weights w_i is on a much faster time scale compared to the time scale α on which the self-averaging quantities R_{ik} and Q_{ij} change.

An appropriate method to deal with such situations is known as adiabatic elimination. It relies on the assumption that the mean value of the fast variable has a stable equilibrium value at each macroscopic time α. One obtains this equilibrium value from the zero of the r.h.s. of (32) with respect to w_i, *i.e.* by investigating the case $dw_i/d\alpha = 0$. The equilibrium values for w_i are thus

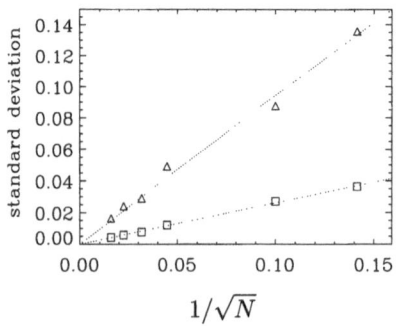
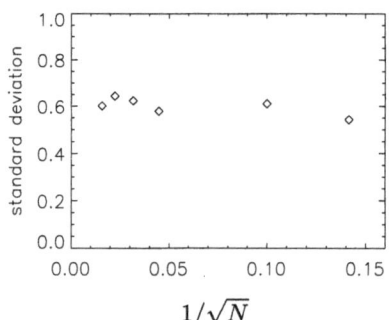

Fig. 3. Finite size analysis of the order parameters R ($\triangle$), Q ($\square$) and w ($\circ$, right panel) in the dynamics (34,35) for the special case $K{=}M{=}1$. Shown are the observed standard deviations as a function of the system size for a fixed value of α. Each point depicts an average taken over 100 simulation runs. As can be seen, R and Q become selfaveraging in the thermodynamic limit $N \to \infty$, *i.e.* their fluctuations vanish in this limit. In contrast to that, the fluctuations of w remain finite if one optimizes the learning rate η_w, which leads to the divergence $\eta_w \to \infty$.

obtained as functions of R_{ik} and Q_{ij} and can be further used to eliminate any appearance of w_i in (30, 31). See [21] for details.

The variational approach discussed in Sec. 2 has also been applied to the analysis of multilayered networks, in particular the soft committee machine [16,26]. Due to the larger number of order parameters the treatment is much more involved than for the simple perceptron network. The investigations show that, in principle, it is possible to reduce the length of the plateau states as discussed above drastically by using appropriate training prescriptions.

4   Dynamics of Prototype Based Learning

In all training scenarios discussed above, the consideration of isotropic, i.i.d. input data yields non-trivial insights, already. The key information is contained in the training labels and, for modeling purposes, we have assumed that they are provided by a teacher network.

In practical situations one would clearly expect the presence of structures in input space, e.g. in the form of clusters which are more or less correlated with the target function. Here we briefly discuss how the theoretical framework has been extended in this direction. We will address, mainly, supervised learning schemes which detect or make use of structures in input space. Unsupervised learning from unlabeled, structured data has been treated along the very same lines but will not be discussed in detail, here. We refer to, for instance, [27,28] for overviews and [34,35,38,43,45] for example results in this context.

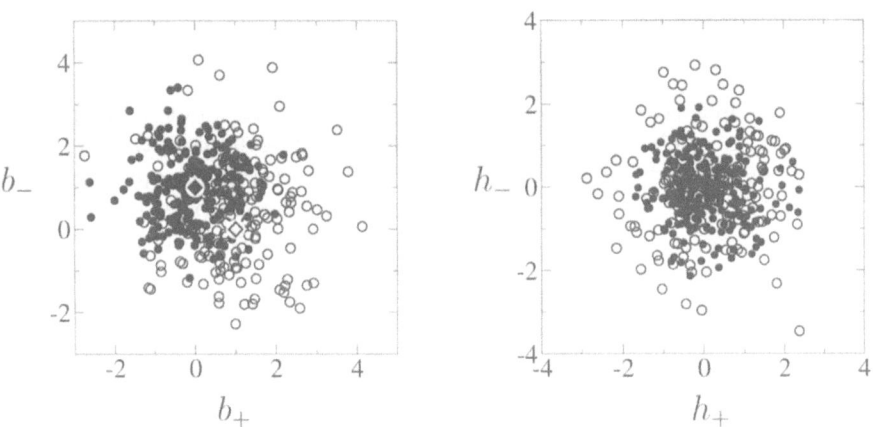

Fig. 4. Data as generated according to the density (36) in $N = 200$ dimensions with example parameters $p_- = 0.6, p_+ = 0.4, v_- = 0.64$, and $v_+ = 1.44$. The open (filled) circles represent $160\,(240)$ vectors $\boldsymbol{\xi}$ from clusters centered about orthonormal vectors ℓB_+ (ℓB_-) with $\ell = 1$, respectively. The left panel displays to the projections $b_\pm = B_\pm \cdot \boldsymbol{\xi}$ and diamonds mark the position of the cluster centers. The right panel shows projections $h_\pm = w_\pm \cdot \boldsymbol{\xi}$ of the same data on a randomly chosen pair of orthogonal unit vectors $w_\pm$.

We will focus on prototype based supervised learning schemes which take into account label information. The popular Learning Vector Quantization algorithm [32] and many of its variants follow the lines of competitive learning. However the aim is to obtain prototypes as typical representatives of their classes which parameterize a distance based classification scheme.

LVQ algorithms can be treated within the same framework as above. The analysis requires only slight modifications due to the assumption of a non-trivial input density.

Several possibilities to model anisotropy in input space have been considered in the literature, a prominent example being unimodal Gaussians with distinct principal axes [29,30,34,35]. Here, we focus on another simple but non-trivial model density: we assume feature vectors are generated independently according to a mixture of isotropic Gaussians:

$$P(\boldsymbol{\xi}) = \sum_{\sigma=\pm 1} p_\sigma P(\boldsymbol{\xi} \mid \sigma) \text{ with } P(\boldsymbol{\xi} \mid \sigma) = \frac{1}{\sqrt{2\pi v_\sigma}^N} \exp\left[-\frac{1}{2v_\sigma}(\boldsymbol{\xi} - \ell \boldsymbol{B}_\sigma)^2\right].$$

(36)

The conditional densities $P(\boldsymbol{\xi} \mid \sigma = \pm 1)$ correspond to clusters with variances v_σ centered at $\ell \boldsymbol{B}_\sigma$. For convenience, we assume that the vectors $\boldsymbol{B}_\sigma$ are orthonormal: $\boldsymbol{B}_\sigma^2 = 1$ and $\boldsymbol{B}_+ \cdot \boldsymbol{B}_- = 0$. The first condition sets only the scale on which the parameter ℓ controls the cluster offset. The orthogonality condition fixes the position of cluster centers with respect to the origin in feature space which could be chosen arbitrarily. Similar densities have been considered in, for instance, [31,37,38,39,40].

In the context of supervised Learning Vector Quantization, see next section, we will assume that the target classification coincides with the cluster membership label σ for each vector $\boldsymbol{\xi}$. Due to the significant overlap of clusters, this task is obviously not linear separable.

Note that linearly separable rules defined for bimodal input data similar to (36) have been studied in [36]. While transient learning curves of the perceptron can differ significantly from the technically simpler case of isotropic inputs, the main results concerning the $(\alpha \to \infty)$–asymptotic behavior persist.

Learning Vector Quantization (LVQ) is a particularly intuitive and powerful family of algorithms which has been applied in a variety of practical problems [41]. LVQ identifies prototypes, i.e. typical representatives of the classes in feature space which then parameterize a distance based classification scheme.

Competitive learning schemes have been suggested in which a set of prototype vectors is updated from a sequence of example data. We will restrict ourselves to the simplest non-trivial case of two prototypes $\boldsymbol{w}_+, \boldsymbol{w}_- \in I\!\!R^N$ and data generated according to a bi-modal distribution of type (36).

Most frequently a *nearest prototype* scheme is implemented: For classification of a given input $\boldsymbol{\xi}$, the distances

$$d_s(\boldsymbol{\xi}) = (\boldsymbol{\xi} - \boldsymbol{w}_s)^2, \quad s = \pm 1 \tag{37}$$

are determined and $\boldsymbol{\xi}$ is assigned to class $+1$ if $d_+(\boldsymbol{\xi}) \leq d_-(\boldsymbol{\xi})$ and to class -1 else. The (squared) Euclidean distance (37) appears to be a natural choice.

In practical situations, however, it can lead to inferior performance and the identification of an appropriate distance or similarity measure is one of the key issues in applications of LVQ.

A simple two prototype system as described above parameterizes a linearly separable classifier, only. However, we will consider learning of a non-separable rule where non-trivial effects of the prototype dynamics can be studied in this simple setting already. Extensions to more complex models with several prototypes, i.e. piecewise linear decision boundaries and multi-modal input densities are possible but non-trivial, see [45] for a recent example.

Generically, LVQ algorithms perform updates of the form

$$w_s^{\mu+1} = w_s^{\mu} + \frac{\eta}{N} f(d_+^{\mu}, d_-^{\mu}, s, \sigma^{\mu}) (\xi^{\mu} - w_s^{\mu}). \tag{38}$$

Hence, prototypes are either moved towards or away from the current input. Here, the modulation function f controls the sign and, together with an explicit learning rate η, the magnitude of the update.

So–called Winner-Takes-All (WTA) schemes update only the prototype which is currently closest to the presented input vector. A prominent example of supervised WTA learning is Kohonen's original formulation, termed LVQ1 [32,33]. In our model scenario it corresponds to Eq. (38) with

$$f(d_+^{\mu}, d_-^{\mu}, s, \sigma^{\mu}) = \Theta(d_{-s}^{\mu} - d_{+s}^{\mu}) s \sigma^{\mu} \tag{39}$$

The Heaviside function singles out the winning prototype, and the product $s \sigma^{\mu} = +1(-1)$ if the labels of prototype and example coincide (disagree).

For the formal analysis of the training dynamics, we can proceed in complete analogy to the previously studied cases of perceptron and layered neural networks. A natural choice of order parameters are the self- and cross-overlaps of the involved N-dimensional vectors:

$$R_{s\sigma} = w_s \cdot B_{\sigma} \quad \text{and} \quad Q_{st} = w_s \cdot w_t \quad \text{with} \quad \sigma, s, t \in \{-1, +1\} \tag{40}$$

While these definitions are formally identical with Eq. (29), the role of the reference vectors B_{σ} is not that of *teacher vectors*, here.

Following the by now familiar lines we obtain a set of coupled ODE of the form

$$\frac{dR_{ST}}{d\alpha} = \eta \left(\langle b_{\tau} f_S \rangle - R_{ST} \langle f_S \rangle \right)$$

$$\frac{dQ_{ST}}{d\alpha} = \eta \left(\langle h_S f_T + h_T f_S \rangle - Q_{ST} \langle f_S + f_T \rangle \right)$$

$$+ \eta^2 \sum_{\sigma=\pm 1} v_{\sigma} p_{\sigma} \langle f_S f_T \rangle_{\sigma}. \tag{41}$$

Here, averages $\langle \ldots \rangle$ over the full density $P(\xi)$, Eq. (36) have to be evaluated as appropriate sums over conditional averages $\langle \ldots \rangle_{\sigma}$ corresponding to ξ drawn from cluster σ:

$$\langle \ldots \rangle = p_+ \langle \ldots \rangle_+ + p_- \langle \ldots \rangle_-.$$

For a large class of LVQ modulation functions, the actual input $\boldsymbol{\xi}^\mu$ appears on the right hand side of Eq. (41) only through its length and the projections

$$h_s = \boldsymbol{w}_s \cdot \boldsymbol{\xi} \quad \text{and} \quad b_\sigma = \boldsymbol{B}_\sigma \cdot \boldsymbol{\xi} \tag{42}$$

where we omitted indices μ but implicitly assume that the input $\boldsymbol{\xi}$ is uncorrelated with the current prototypes $\boldsymbol{w}_s$. Note that also Heaviside terms as in Eq. (39) do not depend on $\boldsymbol{\xi}$ explicitly, for example:

$$\Theta\left(d_- - d_+\right) = \Theta\left[+2(h_+ - h_-) - Q_{++} + Q_{--}\right].$$

When performing the average over the actual example $\boldsymbol{\xi}$ we first exploit the fact that

$$\lim_{N \to \infty} \langle \boldsymbol{\xi}^2 \rangle / N = (v_+ p_+ + v_- p_-)$$

for all input vectors in the thermodynamic limit. Furthermore, the joint Gaussian density $P(h_+^\mu, h_-^\mu, b_+^\mu, b_-^\mu)$ can be expressed as a sum over contributions from the clusters. The respective conditional densities are fully specified by first and second moments

$$\langle h_s \rangle_\sigma = \ell R_{s\sigma}, \quad \langle b_\tau \rangle_\sigma = \ell \delta_{\tau\sigma}, \quad \langle h_s h_t \rangle_\sigma - \langle h_s \rangle_\sigma \langle h_t \rangle_\sigma = v_\sigma Q_{st}$$

$$\langle h_s b_\tau \rangle_\sigma - \langle h_s \rangle_\sigma \langle b_\tau \rangle_\sigma = v_\sigma R_{s\tau}, \quad \langle b_\rho b_\tau \rangle_\sigma - \langle b_\rho \rangle_\sigma \langle b_\tau \rangle_\sigma = v_\sigma \delta_{\rho\tau} \tag{43}$$

where $s, t, \sigma, \rho, \tau \in \{+1, -1\}$ and $\delta_{...}$ is the Kronecker-Delta. Hence, the density of $h_\pm$ and $b_\pm$ is given in terms of the model parameters $\ell, p_\pm, v_\pm$, and the above defined set of order parameters in the previous time step.

After working out the system of ODE for a specific modulation function, it can be integrated, at least numerically. Here we consider prototypes that are initialized as independent random vectors of squared length $\hat{Q}$ with no prior knowledge about the cluster positions. In terms of order parameters this implies in our simple model

$$Q_{++}(0) = Q_{--}(0) = \hat{Q}, \text{ and } Q_{+-}(0) = R_{S\sigma}(0) = 0 \quad \text{for all } S, \sigma. \tag{44}$$

As in any supervised scenario, the success of learning is to be quantified in terms of the generalization error. Here we have to consider two contributions for misclassifying data from cluster $\sigma = 1$ or $\sigma = -1$ separately:

$$\epsilon = p_+ \epsilon_+ + p_- \epsilon_- \quad \text{with} \quad \epsilon_\sigma = \langle \Theta\left(d_{+\sigma} - d_{-\sigma}\right) \rangle_\sigma. \tag{45}$$

Exploiting the central limit theorem in the same fashion as above, one obtains for the above contributions $\epsilon_\pm$:

$$\epsilon_\sigma = \Phi\left(\frac{Q_{\sigma\sigma} - Q_{-\sigma-\sigma} - 2\ell(R_{\sigma\sigma} - R_{-\sigma\sigma})}{2\sqrt{v_\sigma}\sqrt{Q_{++} - 2Q_{+-} + Q_{--}}}\right) \tag{46}$$

where $\Phi(z) = \int_{-\infty}^z dx\, e^{-x^2/2}/\sqrt{2\pi}$.

By inserting $\{R_{S\sigma}(\alpha), Q_{ST}(\alpha)\}$ we obtain the learning curve $\epsilon_g(\alpha)$, i.e. the typical generalization error after on-line training with αN random examples. Here, we once more exploit the fact that the order parameters and, thus, also ϵ_g are self-averaging non-fluctuating quantities in the thermodynamic limit $N \to \infty$.

As an example, we consider the dynamics of LVQ1, cf. Eq. (39). Fig. 5 (left panel) displays the learning curves as obtained for a particular setting of the model parameters and different choices of the learning rate η. Initially, a large learning rate is favorable, whereas smaller values of η facilitate better generalization behavior at large α. One can argue that, as in stochastic gradient descent procedures, the best asymptotic ϵ_g will be achieved for small learning rates $\eta \to 0$. In this limit, we can omit terms quadratic in η from the differential equations and integrate them directly in rescaled time $(\eta\alpha)$. The asymptotic, stationary result for $(\eta\alpha) \to \infty$ then corresponds to the best achievable performance of the algorithm in the given model settings. Figure 5 (right panel) displays, among others, an example result for LVQ1.

With the formalism outlined above it is readily possible to compare different algorithms within the same model situation. This concerns, for instance, the detailed prototype dynamics and sensitivity to initial conditions. Here, we restrict ourselves to three example algorithms and very briefly discuss essential properties and the achievable generalization ability:

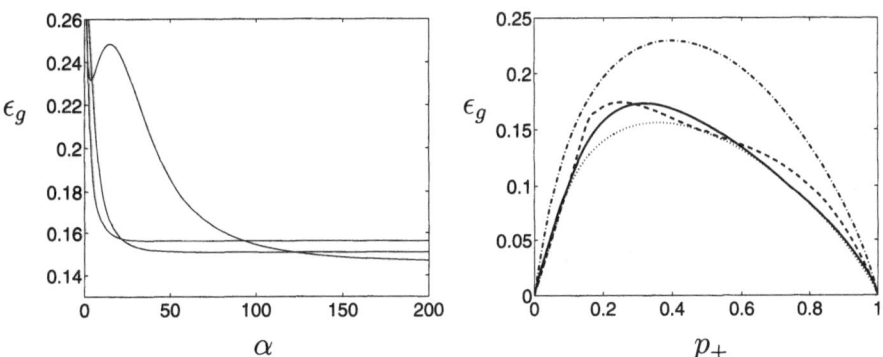

Fig. 5. LVQ learning curves and comparison of algorithms.
Left panel: $\epsilon_g(\alpha)$ for $\ell = 1.2, p_+ = 0.8$, and $v_+ = v_- = 1$. Prototypes were initialized as in Eq. (44) with $\hat{Q} = 10^{-4}$. From bottom to top at $\alpha = 200$, the graphs correspond to learning rates $\eta = 0.2, 1.0$, and 2.0, respectively.
Right panel: Achievable generalization error as a function of $p_+ = 1 - p_-$ in the model with $\ell = 1, v_+ = 0.25$, and $v_- = 0.81$. Initialization of all training schemes as specified in the left panel. The solid line marks the result for LVQ1 in the limits $\eta \to 0$ and $\eta\alpha \to \infty$, the dashed line corresponds to an idealized early stopping procedure applied to LVQ+/-, and the chain line represents the $\alpha \to \infty$ asymptotic outcome of LFM training. In addition, the dotted curve represents the best possible linear decision boundary constructed from the input density.

- **LVQ1:** This basic prescription was already defined above as the original WTA algorithm with modulation function

$$f_s = \Theta(d_{-s} - d_{+s}) \, s \, \sigma.$$

- **LVQ+/-:** Several modifications have been suggested in the literature, aiming at better generalization behavior. The term LVQ+/- is used here to describe one basic variant which updates two vectors at a time: the closest among all prototypes which belong to the same class (a) and the closest one among those that represent a different class (b). The so-called *correct winner* (a) is moved towards the data while the *wrong winner* (b) is pushed even farther away. In our simple model scenario this amounts to the modulation function

$$f_s = s \, \sigma = \pm 1.$$

- **LFM:** The so-called *learning from mistakes* (LFM) scheme performs an update of the LVQ+/- type, but only for inputs which are misclassified before the learning step:

$$f_s = \Theta(d_\sigma - d_{-\sigma}) \, s \, \sigma.$$

The prescription can be obtained as a limiting case of various algorithms suggested in the literature, see [42] for a detailed discussion. Here we emphasize the similarity with the familiar perceptron training discussed in the first sections.

Learning curves and asymptotic behavior of LVQ1 are exemplified in Fig. 5. As an interesting feature one notes a non-monotonicity of $\epsilon_g(\alpha)$ for small learning rates. It corresponds to an over-shooting effect observed in the dynamics of prototypes when approaching their stationary positions [42]. It is important to note that this very basic, heuristic prescription achieves excellent performances: Typically, the asymptotic result is quite close to the optimal ϵ_g, given by the best linear decision boundary as constructed from the input density (36).

The naive application of LVQ+/- training results in a strong instability: In all settings with $p_+ \neq p_-$, the prototype representing the *weaker* class will be pushed away in the majority of training steps. Consequently, a divergent behavior is observed and, for $\alpha \to \infty$, the classifier assigns all data to the stronger cluster with the trivial result $\epsilon_g = \min\{p_+, p_-\}$. Several measures have been suggested to control the instability. In Kohonen's LVQ2.1 and other modifications, example data are only accepted for update when $\boldsymbol{\xi}$ falls into a *window* close to the current decision boundary. Another intuitive approach is based on the observation that $\epsilon_g(\alpha)$ generically displays a pronounced minimum before the generalization behavior deteriorates. In our formal model, it is possible to work out the location of the minimum analytically and thus determine the performance of an idealized *early stopping* method. The corresponding result is displayed in Fig. 5 (right panel, dashed line) and appears to compete well with LVQ1. However, it is important to note that the quality of the minimum in $\epsilon_g(\alpha)$ strongly depends on the initial conditions. Furthermore, in a practical situation, successful early stopping would require the application of costly validation schemes.

Finally, we briefly discuss the LFM prescription. A peculiar feature of LFM is that the stationary position of prototypes does depend strongly on the initial configuration [42]. On the contrary, the $\alpha \to \infty$ asymptotic decision boundary is well-defined. In the LFM prescription, emphasis is on the classification; the aspect of Vector Quantization (representation of clusters) is essentially disregarded. While at first sight clear and attractive, LFM yields a far from optimal performance in the limit $\alpha \to \infty$. Note that already in perceptron training as discussed in the first sections, a naive learning from mistakes strategy is bound to fail miserably in general cases of noisy data or unlearnable rules.

The above considerations concern only the simplest LVQ training scenarios and algorithms. Several directions in which to extend the formalism are obviously interesting.

We only mention that unsupervised prototype based learning has been treated in complete analogy to the above [43,44]. Technically, it reduces to the consideration of modulation functions which do not depend on the cluster or class label. The basic competitive WTA Vector Quantization training would be represented, for instance, by the modulation function $f_s = \Theta(d_{-s} - d_{+s})$ which always moves the winning prototype closer to the data. The training prescription can be interpreted as a stochastic gradient descent of a cost function, the so-called quantization error [44]. The exchange and permutation symmetry of prototypes in unsupervised training results in interesting effects which resemble the plateaus discussed in multilayered neural networks, cf. section 3.

The consideration of a larger number of Gaussians contributing to the input density is relatively simple. Thus, it is possible to model more complex data structures and study their effect on the training dynamics. The treatment of more than two prototypes is also conceptually simple but constitutes a technical challenge. Obviously, the number of order parameters increases. In addition, the r.h.s. of the ODE cannot be evaluated analytically, in general, but involve numerical integrals. Note that the dynamics of several prototypes representing the same class of data resembles strongly the behavior of unsupervised Vector Quantization. First results along these lines have been obtained recently, see for instance [45].

The variational optimization, as discussed for the perceptron in detail, should give insights into the essential features of robust and successful LVQ schemes. Due to the more complex input density, however, the analysis proves quite involved and has not yet been completed.

A highly relevant extension of LVQ is that of relevance learning. Here, the idea is to replace the simple minded Euclidean metrics by an adaptive measure. An important example is a weighted distance of the form

$$d(\boldsymbol{w}, \boldsymbol{\xi}) = \sum_{j=1}^{N} \lambda_j^2 \, (w_j - \xi_j)^2 \quad \text{with} \quad \sum_{j=1}^{N} \lambda_j^2 = 1$$

where the normalized factors λ_j are called relevances as they measure the importance of dimension j in the classification. Relevance LVQ (RLVQ) and related schemes update these factors according to a heuristic or gradient based scheme in the course of training [46,47]. More complex schemes employ a full matrix of relevances in a generalized quadratic measure or consider local measures attached to the prototypes [48].

The analysis of the corresponding training dynamics constitutes another challenge in the theory of on-line learning. The description has to go beyond the techniques discussed in this paper, as the relevances define a time-dependent linear transformation of feature space.

5   Summary and Outlook

The statistical physics approach to learning has allowed for the analytical treatment of the learning dynamics in a large variety of adaptive systems. The consideration of typical properties of large systems in specific model situations complements other approaches and contributes to the theoretical understanding of adaptive systems.

Here, we have highlighted only selected topics as an introduction to this line of research. The approach is presented, first, in terms of the perceptron network. Despite its simplicity, the framework led to highly non-trivial insights and faciliated the putting forward of the method. For instance, the variational approach to optimal training was developed in this context. Gradient based training in multilayered networks constitutes an important example for the analysis of more complex architectures. Here, non-trivial effects such as quasistationary plateau states can be observed and investigated systematically. Finally, a recent application of theoretical framework concerns prototype based training in so-called Learning Vector Quantization.

Several interesting questions and results have not been discussed at all or could be mentioned only very briefly: the study of finite system sizes, on-line learning in networks with discrete weights, unsupervised learning and clustering, training from correlated data or from a fixed pool of examples, query strategies, to name only a few. We can only refer to the list of references, in particular [27] and [28] may serve as a starting point for the interested reader.

Due to the conceptual simplicity of the approach and its applicability in a wide range of contexts it will certainly continue to facilitate better theoretical understanding of learning systems in general. Current challenges include the treatment of non-trivial input distributions, the dynamics of learning in more complex networks architectures, the optimization of algorithms and their practical implementation in such systems, or the investigation of component-wise updates as, for instance, in relevance LVQ.

References

1. Metropolis, N., Rosenbluth, A.W., Rosenbluth, M.N., Teller, A.H., Teller, E.: Equations of State calculations by fast computing machines. J. Chem. Phys. 21, 1087 (1953)
2. Huang, K.: Statistical Mechanics. Wiley and Sons, New York (1987)

3. Jaynes, E.T.: Probability Theory: The Logic of Science. Bretthorst, G.L. (ed.). Cambridge University Press, Cambridge (2003)
4. Mace, C.W.H., Coolen, T.: Dynamics of Supervised Learning with Restricted Training Sets. Statistics and Computing 8, 55–88 (1998)
5. Biehl, M., Schwarze, M.: On-line learning of a time-dependent rule. Europhys. Lett. 20, 733–738 (1992)
6. Biehl, M., Schwarze, H.: Learning drifting concepts with neural networks. Journal of Physics A: Math. Gen. 26, 2651–2665 (1993)
7. Kinouchi, O., Caticha, N.: Lower bounds on generalization errors for drifting rules. J. Phys. A: Math. Gen. 26, 6161–6171 (1993)
8. Vicente, R., Kinouchi, O., Caticha, N.: Statistical Mechanics of Online Learning of Drifting Concepts: A Variational Approach. Machine Learning 32, 179–201 (1998)
9. Reents, G., Urbanczik, R.: Self-averaging and on-line learning. Phys. Rev. Lett. 80, 5445–5448 (1998)
10. Kinzel, W., Rujan, P.: Improving a network generalization ability by selecting examples. Europhys. Lett. 13, 2878 (1990)
11. Kinouchi, O., Caticha, N.: Optimal generalization in perceptrons. J. Phys. A: Math. Gen. 25, 6243–6250 (1992)
12. Copelli, M., Caticha, N.: On-line learning in the committee machine. J. Phys. A: Math. Gen. 28, 1615–1625 (1995)
13. Biehl, M., Riegler, P.: On-line Learning with a Perceptron. Europhys. Lett. 78, 525–530 (1994)
14. Biehl, M., Riegler, P., Stechert, M.: Learning from Noisy Data: An Exactly Solvable Model. Phys. Rev. E 76, R4624–R4627 (1995)
15. Copelli, M., Eichhorn, R., Kinouchi, O., Biehl, M., Simonetti, R., Riegler, P., Caticha, N.: Noise robustness in multilayer neural networks. Europhys. Lett. 37, 427–432 (1995)
16. Vicente, R., Caticha, N.: Functional optimization of online algorithms in multilayer neural networks. J. Phys. A: Math. Gen. 30, L599–L605 (1997)
17. Opper, M.: A Bayesian approach to on-line learning. In: [27], pp. 363–378 (1998)
18. Opper, M., Winther, O.: A mean field approach to Bayes learning in feed-forward neural networks. Phys. Rev. Lett. 76, 1964–1967 (1996)
19. Solla, S.A., Winther, O.: Optimal perceptron learning: an online Bayesian approach. In: [27], pp. 379–398 (1998)
20. Cybenko, G.V.: Approximation by superposition of a sigmoidal function. Math. of Control, Signals and Systems 2, 303–314 (1989)
21. Endres, D., Riegler, P.: Adaptive systems on different time scales. J. Phys. A: Math. Gen. 32, 8655–8663 (1999)
22. Biehl, M., Schwarze, H.: Learning by on-line gradient descent. J. Phys A: Math. Gen. 28, 643 (1995)
23. Saad, D., Solla, S.A.: Exact solution for on-line learning in multilayer neural networks. Phys. Rev. Lett. 74, 4337–4340 (1995)
24. Saad, D., Solla, S.A.: Online learning in soft committee machines. Phys. Rev. E 52, 4225–4243 (1995)
25. Biehl, M., Riegler, P., Wöhler, C.: Transient Dynamics of Online-learning in two-layered neural networks. J. Phys. A: Math. Gen. 29, 4769 (1996)
26. Saad, D., Rattray, M.: Globally optimal parameters for on-line learning in multi-layer neural networks. Phys. Rev. Lett. 79, 2578 (1997)
27. Saad, D. (ed.): On-line learning in neural networks. Cambridge University Press, Cambridge (1998)

28. Engel, A., Van den Broeck, C.: The Statistical Mechanics of Learning. Cambridge University Press, Cambridge (2001)
29. Schlösser, E., Saad, D., Biehl, M.: Optimisation of on-line Principal Component Analysis. J. Physics A: Math. Gen. 32, 4061 (1999)
30. Biehl, M., Schlösser, E.: The dynamics of on-line Principal Component Analysis. J. Physics A: Math. Gen. 31, L97 (1998)
31. Biehl, M., Mietzner, A.: Statistical mechanics of unsupervised learning. Europhys. Lett. 27, 421–426 (1993)
32. Kohonen, T.: Self-Organizing Maps. Springer, Berlin (1997)
33. Kohonen, T.: Learning vector quantization. In: Arbib, M.A. (ed.) The Handbook of Brain Theory and Neural Networks, pp. 537–540. MIT Press, Cambridge (1995)
34. Van den Broeck, C., Reimann, P.: Unsupervised Learning by Examples: On-line Versus Off-line. Phys. Rev. Lett. 76, 2188–2191 (1996)
35. Reimann, P., Van den Broeck, C., Bex, G.J.: A Gaussian Scenario for Unsupervised Learning. J. Phys. A: Math. Gen. 29, 3521–3533 (1996)
36. Riegler, P., Biehl, M., Solla, S.A., Marangi, C.: On-line learning from clustered input examples. In: Marinaro, M., Tagliaferri, R. (eds.) Neural Nets WIRN Vietri 1995, Proc. of the 7th Italian Workshop on Neural Nets, pp. 87–92. World Scientific, Singapore (1996)
37. Marangi, C., Biehl, M., Solla, S.A.: Supervised learning from clustered input examples. Europhys. Lett. 30, 117–122 (1995)
38. Biehl, M.: An exactly solvable model of unsupervised learning. Europhysics Lett. 25, 391–396 (1994)
39. Meir, R.: Empirical risk minimization versus maximum-likelihood estimation: a case study. Neural Computation 7, 144–157 (1995)
40. Barkai, N., Seung, H.S., Sompolinksy, H.: Scaling laws in learning of classification tasks. Phys. Rev. Lett. 70, 3167–3170 (1993)
41. Neural Networks Research Centre. Bibliography on the self-organizing maps (SOM) and learning vector quantization (LVQ). Helsinki University of Technology (2002), http://liinwww.ira.uka.de/bibliography/Neural/SOM.LVQ.html
42. Biehl, M., Ghosh, A., Hammer, B.: Dynamics and generalization ability of LVQ algorithms. J. Machine Learning Research 8, 323–360 (2007)
43. Biehl, M., Freking, A., Reents, G.: Dynamics of on-line competitive learning. Europhysics Letters 38, 73–78 (1997)
44. Biehl, M., Ghosh, A., Hammer, B.: Learning Vector Quantization: The Dynamics of Winner-Takes-All algorithms. Neurocomputing 69, 660–670 (2006)
45. Witeolar, A., Biehl, M., Ghosh, A., Hammer, B.: Learning Dynamics of Neural Gas and Vector Quantization. Neurocomputing 71, 1210–1219 (2008)
46. Bojer, T., Hammer, B., Schunk, D., Tluk von Toschanowitz, K.: Relevance determination in learning vector quantization. In: Verleysen, M. (ed.) European Symposium on Artificial Neural Networks ESANN 2001, pp. 271–276. D-facto publications, Belgium (2001)
47. Hammer, B., Villmann, T.: Generalized relevance learning vector quantization. Neural Networks 15, 1059–1068 (2002)
48. Schneider, P., Biehl, M., Hammer, B.: Relevance Matrices in Learning Vector Quantization. In: Verleysen, M. (ed.) European Symposium on Artificial Neural Networks ESANN 2007, pp. 37–43. d-side publishing, Belgium (2007)

Some Theoretical Aspects of the Neural Gas Vector Quantizer

Thomas Villmann[1], Barbara Hammer[2], and Michael Biehl[3]

[1] University Leipzig, Medical Department,
04103 Leipzig, Germany
thomas.villmann@medizin.uni-leipzig.de
[2] Clausthal University of Technology, Institute of Computer Science,
38678 Clausthal, Germany
hammer@inf.tu-clausthal.de
[3] University of Groningen,
Institute of Mathematics and Computing Science,
P.O. Box 407, 9700 AK Groningen, The Netherlands
m.biehl@rug.nl

Abstract. We investigate the neural gas quantizer in the light of statistical physics concepts. We show that this algorithm can be extended to a vector quantizer with general differentiable similarity measure offering a greater flexibility. Further, we show that the neighborhood cooperativeness control parameter is not equivalent to an inverse temperature like in the deterministic annealing vector quantizer introduced by K. Rose et al. Instead, an annealed variant of neural gas can be obtained using the formalism proposed by T. Heskes for self-organizing maps.

1 Introduction

Neural maps constitute an important neural network paradigm. In brains, neural maps occur in all sensory modalities as well as in motor areas. In technical contexts, neural maps are utilized in the fashion of neighborhood preserving vector quantizers. In both neural maps as well as vector quantization, these networks project data from a *possibly high-dimensional* input space $\mathcal{D} \subseteq \mathbb{R}^d$ onto a set A formally written as $\Psi_{\mathcal{D} \rightarrow A} : \mathcal{D} \rightarrow A$. To achieve this projection, vector quantizers are trained by unsupervised learning schemes. The intimate relationship between neural maps and vector quantization is manifest in the identical projection rule: map a data point to the nearest codebook vector!

In more detail, general vector quantization is an unsupervised method for data compression. This method requires that the data are given in sets of data vectors possibly of rather high dimensions. The data are approximated by the substantially smaller set $\mathbf{W}$ of reference vectors $\mathbf{w}_i$ (prototype vectors), $i \in A$ and a unique mapping between A and $\mathbf{W}$. Then, a data vector $\mathbf{v} \in \mathcal{D}$ is coded by this reference vector $\mathbf{w}_{s(\mathbf{v})}$ the distance $\xi\left(\mathbf{v}, \mathbf{w}_{s(\mathbf{v})}\right)$ of which takes its minimum computed over all elements of $\mathbf{W}$:

$$\Psi_{\mathcal{D} \rightarrow A} : \mathbf{v} \mapsto s(\mathbf{v}) = \underset{i \in A}{\arg\min}\, \xi(\mathbf{v}, \mathbf{w}_i). \tag{1}$$

M. Biehl et al.: (Eds.): Similarity-Based Clustering, LNAI 5400, pp. 23–34, 2009.
© Springer-Verlag Berlin Heidelberg 2009

Frequently, $\xi(\mathbf{v}, \mathbf{w})$ is usual the quadratic Euclidean norm

$$\begin{aligned} \xi(\mathbf{v}, \mathbf{w}) &= \|\mathbf{v} - \mathbf{w}\|^2 \\ &= (\mathbf{v} - \mathbf{w})^2 \end{aligned}$$

Here we only suppose that $\xi(\mathbf{v}, \mathbf{w})$ is a differentiable symmetric similarity measure. In the context of neural vector quantization the elements of A are called neurons, $s(\mathbf{v})$ is called *winner neuron* or *wining unit*. The set

$$\Omega_i = \{\mathbf{v} \in \mathcal{D} | \Psi_{D \to A}(\mathbf{v}) = i\}$$

is called (masked) receptive field of the neuron i.

The crucial point in vector quantization is how one can find a good set of prototypes $\mathbf{W}$. From mathematical point of view an appropriate criterion is the expectation value of the squared expected reconstruction error

$$E_{VQ} = \sum_j \int p(j|\mathbf{v}) \cdot P(\mathbf{v}) \cdot \xi(\mathbf{v}, \mathbf{w}_j) \, d\mathbf{v} \tag{2}$$

with P is the data density and $p(j|\mathbf{v})$ are the crisp assignment variables

$$p(j|\mathbf{v}) = \begin{cases} 1 & \text{iff } j = s(\mathbf{v}) \\ 0 & \text{else} \end{cases} \tag{3}$$

Several adaptation schemes are designed to aim for solving this optimization task efficiently including the well-known algorithms *k-means* [6],[5] or the famous self-organizing map [4]. An approach based on principles of statistical physics is the deterministic annealed vector quantization algorithm [8],[9]. A more efficient method is the neural gas quantizer (NG), which frequently shows a faster convergence [7]. Although both algorithms follow a 'soft-max' strategy it is not clear up to now, whether they are equivalent. This question motivates the paper. Moreover, we consider the NG for more general distance measures than the Euclidean, which would offer a greater flexibility of the algorithm, and, hence, extend the range of applications.

The paper is structured as follows. First, we shortly introduce the NG and show that it is possible to extend them to more general distance measures. After a short review of the deterministic annealed vector quantizer and its basic principles. A comparison with the NG will show fundamental differences such that convergence of NG to the above costfunction (2) can not be derived from this. Therefore, we develop a new strategy for neural gas learning to ensure convergence.

2   The Neural Gas Vector Quantizer

The neural gas vector quantizer (NG) is an unsupervised prototype based vector quantization algorithm in the above sense. During the adaptation process a sequence of data points $\mathbf{v} \in \mathcal{D}$ is presented to the map with respect to the data distribution $P(\mathcal{D})$. Each time the currently most proximate neuron s is determined according to (1), and the prototype $\mathbf{w}_s$ as well as all prototypes $\mathbf{w}_i$ in the neighborhood of $\mathbf{w}_s$ are shifted towards $\mathbf{v}$, according to

$$\triangle \mathbf{w}_j = -\epsilon h_\lambda \left(rk_j \left(\mathbf{v}, \mathbf{W} \right) \right) \frac{\partial \xi \left(\mathbf{v}, \mathbf{w}_j \right)}{\partial \mathbf{w}_j}. \tag{4}$$

The property of "being in the neighborhood of $\mathbf{w}_s$" is captured by the neighborhood function

$$h_\lambda \left(rk_j \left(\mathbf{v}, \mathbf{W} \right) \right) = \exp \left(-\frac{rk_j \left(\mathbf{v}, \mathbf{W} \right)}{\lambda} \right), \tag{5}$$

with the rank function

$$rk_j \left(\mathbf{v}, \mathbf{W} \right) = \sum_i \theta \left(\xi \left(\mathbf{v}, \mathbf{w}_j \right) - \xi \left(\mathbf{v}, \mathbf{w}_i \right) \right) \tag{6}$$

counting the number of pointers $\mathbf{w}_i$ for which the relation $\xi \left(\mathbf{v}, \mathbf{w}_i \right) < \xi \left(\mathbf{v}, \mathbf{w}_j \right)$ is valid [7]. Here, $\theta \left(x \right)$ is the Heaviside-function. We remark that the neighborhood function is evaluated in the input space $\mathcal{D}$. In case of $\xi \left(\mathbf{v}, \mathbf{w} \right)$ being the quadratic Euclidean norm MARTINETZ has shown that the above given adaptation rule for the weight vectors follows in average a gradient dynamic according to the potential function [7]:

$$E_{NG} = \frac{1}{2C \left(\lambda \right)} \sum_j \int P \left(\mathbf{v} \right) h_\lambda \left(rk_j \left(\mathbf{v}, \mathbf{W} \right) \right) \xi \left(\mathbf{v}, \mathbf{w}_j \right) d\mathbf{v} \tag{7}$$

with $C \left(\lambda \right)$ is a λ-dependent constant

$$C \left(\lambda \right) = \sum_k \exp \left(-\frac{k}{\lambda} \right). \tag{8}$$

It was shown in many applications that the NG shows a robust behavior together with a high precision of learning. For small but non-vanishing neighborhood λ the distribution density ρ of the prototypes follows a power law (magnification law)

$$\rho \left(\mathbf{w}_i \right) \sim P \left(\mathbf{w}_i \right)^{\alpha_{NG}}$$

with the magnification factor (exponent)

$$\alpha_{NG} = \frac{d_{\text{Haussdorff}}}{d_{\text{Haussdorff}} + 2}$$

and $d_{\text{Haussdorff}}$ is the Hausdorff-dimension of the data space $\mathcal{D}$.

The cost function (7) remains valid, if other distance measures than the quadratic Euclidean are used as it is shown in the appendix 1. Obviously, the change of the distance measure changes the properties of the NG and, hence, influences the achieved prototype distribution. For example, concave/convex learning uses the similarity measure

$$\xi_{\varkappa}(\mathbf{v}, \mathbf{w}) = (\mathbf{v} - \mathbf{w})^{\varkappa}$$
$$\overset{def}{=} (\mathbf{v} - \mathbf{w}) \|\mathbf{v} - \mathbf{w}\|^{\varkappa - 1}$$

which leads to the magnification factor

$$\alpha_{NG}^{concave/convex} = \frac{d}{d + 1 + \varkappa}$$

as it was demonstrated in [11].

3 Statistical Physics Interpretation of (Unsupervised) Vector Quantization and Neural Gas

3.1 Vector Quantizer by Deterministic Annealing (VQDA)

ROSE ET AL. introduced an algorithm for vector quantization based on an annealing strategy which can be described in terms of statistical physics to study its behavior [8],[9]. We briefly outline the main sketch of this scheme to have the basis for investigation of the NG in a statistical physics framework.

Following ROSE ET AL. we consider the general cost function for vector quantization (2) with the similarity measure usually taken as $\xi\left(\mathbf{v}, \mathbf{w}_{s(\mathbf{v})}\right) = \left(\mathbf{v} - \mathbf{w}_{s(\mathbf{v})}\right)^2$ the squared Euclidean distance and $s(\mathbf{v})$ as in (1). However, in difference to the the cost function (2) the assignments $p(j|\mathbf{v})$ are here taken as Gibbs distributions

$$p_{\beta}(j|\mathbf{v}) = \frac{\exp\left(-\beta \cdot \xi\left(\mathbf{v}, \mathbf{w}_j\right)\right)}{Z_{\mathbf{v}}}. \tag{9}$$

The values $p_{\beta}(j|\mathbf{v})$ are interpreted as soft assignments and become crisp in the limit $\beta \to \infty$. Correspondingly, in this case each data vector is uniquely assigned (crisp) to a prototype as given in (3).

According to the statistical physics framework the partition function

$$Z_{\mathbf{v}}(\beta) = \sum_k \exp\left(-\beta \cdot \xi(\mathbf{v}, \mathbf{w}_k)\right) \tag{10}$$

can be defined. Then, the parameter $\beta = \frac{1}{T}$ can be seen as a Lagrangian multiplier determined by the given value of E_{VQ}. It is inversely proportional to the temperature $T = \frac{1}{\beta}$ in physics analogy. For a given set $\mathbf{W}$ of prototypes it is assumed that the probability assignments are independent such that the total partition function can be written as the product

$$Z(\beta) = \Pi_{\mathbf{v}} Z_{\mathbf{v}}(\beta). \tag{11}$$

Applying the principle of maximum entropy the corresponding *free energy functional* is

$$F_{VQ}(\beta) = -\ln Z(\beta) \tag{12}$$

$$= -\frac{1}{\beta} \sum_{\mathbf{v}} \ln \left(\sum_j \exp\left(-\beta \xi(\mathbf{v}, \mathbf{w}_j)\right) \right) \tag{13}$$

$$= -\frac{1}{\beta} \int \ln \left[\sum_j \exp\left(-\beta \xi\left(\mathbf{v}, \mathbf{w}_j\right)\right) \right] \cdot P\left(\mathbf{v}\right) d\mathbf{v} \tag{14}$$

where (14) is the continuos variant of (13). We emphasize here for later argumentation that the free energy, on the one hand side, depends on the temperature T by the pre-factor of the integral. On the other hand, the temperature occurs within the exponential term, whereas the original similarity measure ξ remains independent from T.

It was shown by GEMAN&GEMAN that the respective deterministic annealing scheme leads to the convergence to the error functional (2) [1].

3.2   Neural Gas Interpretations

Original NG. We now consider the cost function (7) of NG in the light of this statistical physics frame work. T. MARTINETZ suggested in [7] to take the neighborhood cooperativeness parameter λ, defining the range of interactions in NG, as an inverse temperature in the statistical physics framework. Following this suggestion, we can try to interpret formally the cost function in terms of a free energy description. This gives the motivation of the following considerations.

To compare the NG cost function with the deterministic annealing framework, we assume that in (7) the summation and the integral can be interchanged (LEMMA FUBINI, [10]):

$$\sum_j \int P\left(\mathbf{v}\right) \cdot \nu\left(j|\mathbf{v}\right) \cdot \xi\left(\mathbf{v}, \mathbf{w}_j\right) d\mathbf{v} = \int \left[\sum_j \nu_\lambda\left(j|\mathbf{v}\right) \xi\left(\mathbf{v}, \mathbf{w}_j\right) \right] \cdot P\left(\mathbf{v}\right) d\mathbf{v} \tag{15}$$

with the new assignment variables

$$\nu_\lambda\left(j|\mathbf{v}\right) = \frac{h_\lambda\left(rk_j\left(\mathbf{v}, \mathbf{W}\right)\right)}{C\left(\lambda\right)} \tag{16}$$

$$= \frac{\exp\left(-\frac{rk_j\left(\mathbf{v}, \mathbf{W}\right)}{\lambda}\right)}{C\left(\lambda\right)} \tag{17}$$

We observe that the new variables $\nu_\lambda\left(j|\mathbf{v}\right)$ differ from the assignments $p_\beta\left(j|\mathbf{v}\right)$ in (9): the assignments $p_\beta\left(j|\mathbf{v}\right)$ directly depend on the similarity measure $\xi\left(\mathbf{v}, \mathbf{w}_k\right)$ whereas the dependence in case of the $\nu_\lambda\left(j|\mathbf{v}\right)$ is more complicate:

$$\nu_\lambda\left(j|\mathbf{v}\right) = \frac{\exp\left(-\frac{\sum_i \theta\left(\xi\left(\mathbf{v}, \mathbf{w}_j\right) - \xi\left(\mathbf{v}, \mathbf{w}_i\right)\right)}{\lambda}\right)}{C\left(\lambda\right)} \tag{18}$$

From this it is clear that the cost function of NG can not be seen as a free energy of any cost function because of this structural differences and, therefore, the neighborhood range parameter λ formally does not play the same role as the inverse temperature in the above deterministic annealing approach. Hence, the suggestion given by T. MARTINETZ in [7] is misleading.

Otherwise, the statistical physics interpretation for the VQDA guaranties the convergence for the low-temperature-scenario in the limit to the optimum vector quantizer minimizing the error functional (2). In the following we will apply another framework to a modified variant of NG. This framework was proposed by T. HESKES to modify the original self-organizing map algorithm (SOM) [3] in such a way that for each value of neighborhood range a thermodynamic interpretation is possible and the obtained algorithm follows the respective cost functional.

Modified NG. In this chapter we follow the free energy approach applied to SOM presented by T. HESKES [3] to obtain a statistical physics description for the NG for each neighborhood range λ. For this purpose, we define the local cost for each prototype as

$$lc_j \left(\mathbf{v}, \mathbf{W}\right) = \frac{h_\lambda \left(rk_j \left(\mathbf{v}, \mathbf{W}\right)\right) \cdot \xi \left(\mathbf{v}, \mathbf{w}_j\right)}{2C \left(\lambda\right)} \tag{19}$$

such that the cost function for a single input $\mathbf{v}$ reads as

$$E \left(\mathbf{W}, \mathbf{p_v}, \mathbf{v}\right) = \sum_j p_{\mathbf{v},j} \cdot lc_j \left(\mathbf{v}, \mathbf{W}\right) \tag{20}$$

whereby $p_{\mathbf{v},j} \in [0,1]$ are assignment variables subject to the constraint $\sum_j p_{\mathbf{v},j} = 1$, collected in the vector $\mathbf{p_v}$.

Let s be the index of the winning prototype, but now determined according to the local costs

$$s \left(\mathbf{v}\right) = \underset{j \in A}{\operatorname{argmin}} \, lc_j \left(\mathbf{v}, \mathbf{W}\right) \tag{21}$$

which is different from the winner determination (1) in usual NG. Then the choice for minimizing $E \left(\mathbf{W}, \mathbf{p_v}, \mathbf{v}\right)$ in (20) for a fixed prototype set and given data $\mathbf{v}$ is the crisp assignment $p_{\mathbf{v},s(\mathbf{v})} = 1$ and 0 elsewhere. Defining formally the free energy functional we get

$$F \left(\mathbf{W}, \mathbf{p_v}, \mathbf{v}\right) = E \left(\mathbf{W}, \mathbf{p_v}, \mathbf{v}\right) - T \cdot H \left(\mathbf{p_v}\right) \tag{22}$$

with now softened assignments $p_{\mathbf{v},j} \geq 0$ but preserving the restriction $\sum_j p_{\mathbf{v},j} = 1$. The vale

$$H \left(\mathbf{p_v}\right) = - \sum_j p_{\mathbf{v},j} \cdot \ln \left(p_{\mathbf{v},j}\right) \tag{23}$$

is the Shannon entropy. In this scenario, for a given temperature $T = \frac{1}{\beta}$ and fixed $\mathbf{W}$ and $\mathbf{v}$, the optimum *soft* assignments are given by

$$p_{\mathbf{v},j}|\mathbf{w} = \frac{\exp \left(-\beta \cdot lc_j \left(\mathbf{v}, \mathbf{W}\right)\right)}{\sum_k \exp \left(-\beta \cdot lc_k \left(\mathbf{v}, \mathbf{W}\right)\right)} \tag{24}$$

and the respective minimal free energy reads as

$$F \left(\mathbf{W}, \mathbf{v}\right) = -\frac{1}{\beta} \ln \left[\sum_k \exp \left(-\beta \cdot lc_k \left(\mathbf{v}, \mathbf{W}\right)\right)\right]. \tag{25}$$

Then the online learning for a given temperature T follows a stochastic gradient descent of the free energy according to

$$\triangle \mathbf{w}_j = \frac{\partial F (\mathbf{W}, \mathbf{v})}{\partial \mathbf{w}_j} \tag{26}$$

which is obtained (see Appendix 2) as

$$\triangle \mathbf{w}_j = \frac{p_{\mathbf{v},j}|\mathbf{w}}{2C(\lambda)} \cdot h_\lambda (rk_j (\mathbf{v}, \mathbf{W})) \cdot \frac{\partial \xi (\mathbf{v}, \mathbf{w}_j)}{\partial \mathbf{w}_j} \tag{27}$$

Hence, an adiabtic decreasing of the neighborhood λ during this learning would lead to the global minimum of the vector quantization cost function (2), at least in theory.

In the next steps we concentrate on the batch mode of NG. Following the argumentation in [3], the free energy for a *new* state $\mathbf{W}_{new}$ and a single input $\mathbf{v}$ can be expressed as

$$F (\mathbf{W}_{new}, \mathbf{v}) = lc_j (\mathbf{v}, \mathbf{W}_{new}) + \frac{1}{\beta} \ln (p_{\mathbf{v},j}|\mathbf{w}_{new}) \tag{28}$$

$$= \sum_j p_{\mathbf{v},j}|\mathbf{w}_{old} \cdot lc_j (\mathbf{v}, \mathbf{W}_{new}) + \frac{1}{\beta} \sum_j p_{\mathbf{v},j}|\mathbf{w}_{old} \cdot \ln (p_{\mathbf{v},j}|\mathbf{w}_{new}) \tag{29}$$

$$= \tilde{E} (\mathbf{W}_{new}, \mathbf{p}_{\mathbf{v}}|\mathbf{w}_{old}, \mathbf{v}) - T \cdot \tilde{H} (\mathbf{p}_{\mathbf{v}}|\mathbf{W}_{new}, \mathbf{W}_{old}) \tag{30}$$

whereby the middle expression is obtained from the first one by multiplying with $p_{\mathbf{v},j}|\mathbf{w}_{old}$ and summation over j. The last equation shows the correspondence to (22), however, (30) holds for any choice of $p_{\mathbf{v},j}|\mathbf{w}_{old}$ subject to the constraint that $p_{\mathbf{v},j}|\mathbf{w}_{new}$ is in the form of (24) deduced from minimization of (22).

The term $\sum_j \cdot p_{\mathbf{v},j}|\mathbf{w}_{old} \cdot \ln (p_{\mathbf{v},j}|\mathbf{w}_{new})$ is maximized for $\mathbf{W}_{new} = \mathbf{W}_{old}$; any other choice is at least as good. Thus, minimization of (30) by an EM-scheme only requires a reduction of the first term $\tilde{E} (\mathbf{W}_{new}, \mathbf{p}_{\mathbf{v}}|\mathbf{w}_{old}, \mathbf{v}) = \sum_j \cdot p_{\mathbf{v},j}|\mathbf{w}_{old} \cdot lc_j (\mathbf{v}, \mathbf{W}_{new})$.

The E-step of the batch variant (finite data set is assumed) requires the determination of the assignments $p_{\mathbf{v},j} = p_{\mathbf{v},j}|\mathbf{w}_{old}$ according to (24) for the current state and the prototypes are updated in the subsequent M-step by

$$\mathbf{w}_i = \frac{\sum_{\mathbf{v}} p_{\mathbf{v},i} \cdot lc_i (\mathbf{v}, \mathbf{W})}{\sum_{\mathbf{v}} p_{\mathbf{v},i}} \tag{31}$$

$$= \frac{\sum_{\mathbf{v}} p_{\mathbf{v},i} \cdot lc_i (\mathbf{v}, \mathbf{W})}{\sum_{\mathbf{v}} p_{\mathbf{v},i}} \tag{32}$$

Obviously, this procedure depends only weakly on the concrete choice of the similarity measure $\xi (\mathbf{v}, \mathbf{w}_j)$. Recently, it has been shown that, if the existence of a maybe unknown data metric can be assumed, which is implicitly given by the relations between the finite data examples, the neural gas batch-learning scheme can be adopted with only a few small changes – relational NG [2].

4   Concluding Remarks

In this contribution we outlined some theoretical aspects of the neural gas vector quantizer. In particular we showed that the approach can be extended to a vector quantization algorithm with general differentiable similarity measure, which offers greater flexibility and broader range of applications. Further we demonstrated that the neighborhood cooperativeness parameter is not equivalent to an inverse temperature like in the deterministic annealing vector quantizer approach. However, applying the same framework as done for the self-organizing map by T. Heskes, an annealed variant of NG can be developed such that for adiabatic shrinking of the neighborhood the global minimum of the vector quantization cost functional could be obtained at least in theory. This, however, changes the determination of the mapping rule slightly by incorporation of the, in the limit vanishing neighborhood cooperativeness. Finally, we have to say that this paper leaves the question of convergence in the limit of vanishing neighborhood cooperativeness of the *original* neural gas still as an open problem.

References

1. Geman, S., Geman, D.: Stochastic relaxation, Gibbs distributions, and the Bayesian restoration of images. IEEE Transactions on Pattern Analysis and Machine Intelligence 6, 721–741 (1984)
2. Hammer, B., Hasenfuss, A.: Relational topographic maps. Ifi 07-01, Clausthal University of Technology, Clausthal, Germany (2007)
3. Heskes, T.: Energy functions for self-organizing maps. In: Oja, E., Kaski, S. (eds.) Kohonen Maps, pp. 303–316. Elsevier, Amsterdam (1999)
4. Kohonen, T.: Self-Organizing Maps. Springer Series in Information Sciences, vol. 30. Springer, Heidelberg (1995) (Second Extended Edition 1997)
5. Linde, Y., Buzo, A., Gray, R.: An algorithm for vector quantizer design. IEEE Transactions on Communications 28, 84–95 (1980)
6. MacQueen, J.: Some methods for classification and analysis of multivariate observations. In: LeCam, L., Neyman, J. (eds.) Proceedings of the Fifth Berkeley Symposium on Mathematics, Statistics, and Probability, pp. 281–297. University of California Press, Berkeley (1967)
7. Martinetz, T.M., Berkovich, S.G., Schulten, K.J.: 'Neural-gas' network for vector quantization and its application to time-series prediction. IEEE Trans. on Neural Networks 4(4), 558–569 (1993)
8. Rose, K., Gurewitz, E., Fox, G.: Statistical mechanics and phase transitions in clustering. Physical Review Letters 65(8), 945–948 (1990)
9. Rose, K., Gurewitz, E., Fox, G.: Vector quantization by deterministic annealing. IEEE Transactions on Information Theory 38(4), 1249–1257 (1992)
10. Triebel, H.: Analysis und mathematische Physik, 3rd revised edn. BSB B.G. Teubner Verlagsgesellschaft, Leipzig (1989)
11. Villmann, T., Claussen, J.-C.: Magnification control in self-organizing maps and neural gas. Neural Computation 18(2), 446–469 (2006)

Appendix 1

We have to proof in this appendix that the NG-adaptation rule (4)

$$\triangle \mathbf{w}_i = -\epsilon h_\lambda \left(\mathbf{v}, \mathbf{W}, i \right) \frac{\partial \xi \left(\mathbf{v}, \mathbf{w}_i \right)}{\partial \mathbf{w}_i}$$

for general differentiable, symmetric similarity measure performs a stochastic gradient descent on the cost function (7)

$$E_{NG} = \frac{1}{2C\left(\lambda\right)} \sum_j \int P\left(\mathbf{v}\right) h_\lambda \left(rk_j \left(\mathbf{v}, \mathbf{W} \right) \right) \xi \left(\mathbf{v}, \mathbf{w}_j \right) d\mathbf{v}.$$

For this purpose we consider the gradient $\frac{\partial E_{NG}}{\partial \mathbf{w}_k}$ and obtain

$$\frac{\partial E_{NG}}{\partial \mathbf{w}_k} = \frac{\partial \left[\int \sum_j P\left(\mathbf{v}\right) h_\lambda \left(rk_j \left(\mathbf{v}, \mathbf{W} \right) \right) \xi \left(\mathbf{v}, \mathbf{w}_j \right) d\mathbf{v} \right]}{\partial \mathbf{w}_k}$$

$$= \int \sum_j P\left(\mathbf{v}\right) h_\lambda \left(rk_j \left(\mathbf{v}, \mathbf{W} \right) \right) \frac{\partial \xi \left(\mathbf{v}, \mathbf{w}_j \right)}{\partial \mathbf{w}_k} d\mathbf{v} + \qquad (33)$$

$$\underbrace{\int \sum_j P\left(\mathbf{v}\right) \frac{\partial h_\lambda \left(rk_j \left(\mathbf{v}, \mathbf{W} \right) \right)}{\partial \mathbf{w}_k} \xi \left(\mathbf{v}, \mathbf{w}_j \right) d\mathbf{v}}_{=T_2} \qquad (34)$$

using the FUBINI-LEMMA to interchange integration and summation and dropping the constant $\frac{1}{2C(\lambda)}$. The first term (33) yields the desired learning rule. Hence, it remains to show that the second term T_2 (34) vanishes.

We consider the derivative $\frac{rk_j(\mathbf{v},\mathbf{W})}{\partial \mathbf{w}_k}$: Using the definition (6) of the rank function $rk_j \left(\mathbf{v}, \mathbf{W} \right)$ we obtain

$$\frac{rk_j \left(\mathbf{v}, \mathbf{W} \right)}{\partial \mathbf{w}_k} = \sum_i \frac{\partial \theta \left(\xi \left(\mathbf{v}, \mathbf{w}_j \right) - \xi \left(\mathbf{v}, \mathbf{w}_i \right) \right)}{\partial \mathbf{w}_k}$$

$$= \sum_i \delta_{ji} \left(\mathbf{v} \right) \left(\frac{\partial \xi \left(\mathbf{v}, \mathbf{w}_j \right)}{\partial \mathbf{w}_k} - \frac{\partial \xi \left(\mathbf{v}, \mathbf{w}_i \right)}{\partial \mathbf{w}_k} \right)$$

with

$$\delta_{ji} \left(\mathbf{v} \right) = \delta \left(\xi \left(\mathbf{v}, \mathbf{w}_j \right) - \xi \left(\mathbf{v}, \mathbf{w}_i \right) \right) \qquad (35)$$

being the Dirac-functional. This leads to

$$T_2 = \int \sum_j P\left(\mathbf{v}\right) \frac{\partial h_\lambda \left(rk_j \left(\mathbf{v}, \mathbf{W} \right) \right)}{\partial \mathbf{w}_k} \xi \left(\mathbf{v}, \mathbf{w}_j \right) d\mathbf{v}$$

$$= \int \sum_j P\left(\mathbf{v}\right) \cdot h'_\lambda \cdot \frac{rk_j \left(\mathbf{v}, \mathbf{W} \right)}{\partial \mathbf{w}_k} \cdot \xi \left(\mathbf{v}, \mathbf{w}_j \right) d\mathbf{v}$$

$$= \int \sum_j P\left(\mathbf{v}\right) \cdot h'_\lambda \left(rk_j\left(\mathbf{v}, \mathbf{W}\right)\right) \cdot \xi\left(\mathbf{v}, \mathbf{w}_j\right) \cdot$$

$$\cdot \sum_i \delta_{ji}\left(\mathbf{v}\right) \left(\frac{\partial \xi\left(\mathbf{v}, \mathbf{w}_j\right)}{\partial \mathbf{w}_k} - \frac{\partial \xi\left(\mathbf{v}, \mathbf{w}_i\right)}{\partial \mathbf{w}_k}\right) d\mathbf{v}$$

$$= \int \sum_j P\left(\mathbf{v}\right) \cdot h'_\lambda \left(rk_j\left(\mathbf{v}, \mathbf{W}\right)\right) \cdot \xi\left(\mathbf{v}, \mathbf{w}_j\right) \cdot \frac{\partial \xi\left(\mathbf{v}, \mathbf{w}_j\right)}{\partial \mathbf{w}_k} \cdot \sum_i \delta_{ji}\left(\mathbf{v}\right) d\mathbf{v}$$

$$- \int \sum_j P\left(\mathbf{v}\right) \cdot h'_\lambda \left(rk_j\left(\mathbf{v}, \mathbf{W}\right)\right) \cdot \xi\left(\mathbf{v}, \mathbf{w}_j\right) \cdot \sum_i \delta_{ji}\left(\mathbf{v}\right) \frac{\partial \xi\left(\mathbf{v}, \mathbf{w}_i\right)}{\partial \mathbf{w}_k} d\mathbf{v}$$

$$= \int \sum_i P\left(\mathbf{v}\right) \cdot h'_\lambda \left(rk_i\left(\mathbf{v}, \mathbf{W}\right)\right) \cdot \xi\left(\mathbf{v}, \mathbf{w}_k\right) \cdot \frac{\partial \xi\left(\mathbf{v}, \mathbf{w}_k\right)}{\partial \mathbf{w}_k} \cdot \delta_{ki}\left(\mathbf{v}\right) d\mathbf{v}$$

$$- \int \sum_j P\left(\mathbf{v}\right) \cdot h'_\lambda \left(rk_j\left(\mathbf{v}, \mathbf{W}\right)\right) \cdot \xi\left(\mathbf{v}, \mathbf{w}_j\right) \cdot \delta_{jk}\left(\mathbf{v}\right) \frac{\partial \xi\left(\mathbf{v}, \mathbf{w}_k\right)}{\partial \mathbf{w}_k} d\mathbf{v}$$

The Dirac-functional δ only contributes if $\xi\left(\mathbf{v}, \mathbf{w}_k\right) = \xi\left(\mathbf{v}, \mathbf{w}_i\right) = \xi\left(\mathbf{v}, \mathbf{w}_j\right)$. For that case, however, $rk_j\left(\mathbf{v}, \mathbf{W}\right) = rk_i\left(\mathbf{v}, \mathbf{W}\right)$ and, therefore, $h'_\lambda\left(rk_j\left(\mathbf{v}, \mathbf{W}\right)\right) = h'_\lambda\left(rk_i\left(\mathbf{v}, \mathbf{W}\right)\right)$. Using the symmetry of the Dirac functional (35) both integrals are identical and, hence, their difference vanishes. This conpletes the proof.

Appendix 2

We derive in this appendix the online-learning rule for the modified NG from sec. 3.2. We will show that the update is

$$\triangle \mathbf{w}_j = \frac{p_{\mathbf{v},j|\mathbf{w}}}{2C\left(\lambda\right)} \cdot h_\lambda\left(rk_j\left(\mathbf{v}, \mathbf{W}\right)\right) \cdot \frac{\partial \xi\left(\mathbf{v}, \mathbf{w}_j\right)}{\partial \mathbf{w}_j} \tag{36}$$

as stated in (27). We start with the formal derivative of the respective free energy (26)

$$\triangle \mathbf{w}_j = \frac{\partial F\left(\mathbf{W}, \mathbf{v}\right)}{\partial \mathbf{w}_j} \tag{37}$$

$$= \frac{\partial - \frac{1}{\beta} \ln\left[\sum_i \exp\left(-\beta \cdot lc_i\left(\mathbf{v}, \mathbf{W}\right)\right)\right]}{\partial \mathbf{w}_j} \tag{38}$$

$$= -\frac{1}{\beta} \frac{1}{\sum_i \exp\left(-\beta \cdot lc_i\left(\mathbf{v}, \mathbf{W}\right)\right)} \cdot \sum_i \frac{\partial \exp\left(-\beta \cdot lc_i\left(\mathbf{v}, \mathbf{W}\right)\right)}{\partial \mathbf{w}_j} \tag{39}$$

$$= \sum_i p_{\mathbf{v},i|\mathbf{w}} \cdot \frac{\partial lc_i\left(\mathbf{v}, \mathbf{W}\right)}{\partial \mathbf{w}_j} \tag{40}$$

using the definition (24) of the soft-assignments $p_{\mathbf{v},i|\mathbf{w}}$. We consider the the local cost $lc_i\left(\mathbf{v}, \mathbf{W}\right)$ from (19) or equivalently $f_i\left(\mathbf{v}, \mathbf{W}\right) = 2C\left(\lambda\right) \cdot lc_i\left(\mathbf{v}, \mathbf{W}\right)$ and obtain for the derivate $\frac{\partial f_i(\mathbf{v}, \mathbf{W})}{\partial \mathbf{w}_j}$

$$\frac{\partial f_i(\mathbf{v}, \mathbf{W})}{\partial \mathbf{w}_j} = \frac{\partial \left[h_\lambda \left(rk_i(\mathbf{v}, \mathbf{W})\right) \cdot \xi(\mathbf{v}, \mathbf{w}_i)\right]}{\partial \mathbf{w}_j} \tag{41}$$

$$= \xi(\mathbf{v}, \mathbf{w}_i) \frac{\partial \left[h_\lambda \left(rk_i(\mathbf{v}, \mathbf{W})\right)\right]}{\partial \mathbf{w}_j} + \frac{\partial \xi(\mathbf{v}, \mathbf{w}_i)}{\partial \mathbf{w}_j} h_\lambda \left(rk_i(\mathbf{v}, \mathbf{W})\right) \tag{42}$$

$$= h_\lambda \left(rk_i(\mathbf{v}, \mathbf{W})\right) \left(\frac{\partial \xi(\mathbf{v}, \mathbf{w}_i)}{\partial \mathbf{w}_j} - \frac{1}{\lambda} \xi(\mathbf{v}, \mathbf{w}_i) \cdot \frac{\partial \left[rk_i(\mathbf{v}, \mathbf{W})\right]}{\partial \mathbf{w}_j}\right) \tag{43}$$

using

$$\frac{\partial \left[h_\lambda \left(rk_i(\mathbf{v}, \mathbf{W})\right)\right]}{\partial \mathbf{w}_j} = -\frac{1}{\lambda} h_\lambda \left(rk_i(\mathbf{v}, \mathbf{W})\right) \cdot \frac{\partial \left[rk_i(\mathbf{v}, \mathbf{W})\right]}{\partial \mathbf{w}_j} \tag{44}$$

Hence, equation (40) can be rewritten as

$$\triangle \mathbf{w}_j = \frac{1}{2C(\lambda)} \cdot \sum_i p_{\mathbf{v},i}|\mathbf{w} \cdot h_\lambda \left(rk_i(\mathbf{v}, \mathbf{W})\right) \frac{\partial \xi(\mathbf{v}, \mathbf{w}_i)}{\partial \mathbf{w}_j} - R_i \tag{45}$$

with

$$R_i = \frac{1}{2C(\lambda)\lambda} \cdot \sum_i p_{\mathbf{v},i}|\mathbf{w} \cdot h_\lambda \left(rk_i(\mathbf{v}, \mathbf{W})\right) \cdot \xi(\mathbf{v}, \mathbf{w}_i) \cdot \frac{\partial \left[rk_i(\mathbf{v}, \mathbf{W})\right]}{\partial \mathbf{w}_j} \tag{46}$$

The first term in the update rule can be further simplified using $\frac{\partial \xi(\mathbf{v}, \mathbf{w}_i)}{\partial \mathbf{w}_j} = 0$ for $i \neq j$:

$$\triangle \mathbf{w}_j = \frac{p_{\mathbf{v},j}|\mathbf{w}}{2C(\lambda)} \cdot h_\lambda \left(rk_j(\mathbf{v}, \mathbf{W})\right) \cdot \frac{\partial \xi(\mathbf{v}, \mathbf{w}_j)}{\partial \mathbf{w}_j} - R_i \tag{47}$$

In order to prove (36) it remains to show that $\tilde{R}_i = 2C(\lambda)\lambda \cdot R_i$ vanishes. For this purpose we consider

$$\frac{\partial \left[rk_i(\mathbf{v}, \mathbf{W})\right]}{\partial \mathbf{w}_j} = \frac{\partial \left[\sum_l \theta \left(\xi(\mathbf{v}, \mathbf{w}_i) - \xi(\mathbf{v}, \mathbf{w}_l)\right)\right]}{\partial \mathbf{w}_j} \tag{48}$$

$$= \hat{\delta}_{i,j} \cdot \left(\frac{\partial \xi(\mathbf{v}, \mathbf{w}_i)}{\partial \mathbf{w}_j} \cdot \sum_l \delta_{il}(\mathbf{v})\right) \tag{49}$$

$$- \frac{\partial \xi(\mathbf{v}, \mathbf{w}_j)}{\partial \mathbf{w}_j} \cdot \delta_{ij}(\mathbf{v}) \tag{50}$$

whereby $\hat{\delta}_{i,j}$ is the Kronecker symbol which should not be confused with the Dirac functional $\delta_{ij}(\mathbf{v})$ from (35). Hence, we get

$$\tilde{R}_i = \sum_i p_{\mathbf{v},i} | \mathbf{w} \cdot \left[\xi\left(\mathbf{v}, \mathbf{w}_i\right) h_\lambda\left(rk_i\left(\mathbf{v}, \mathbf{W}\right)\right) \cdot \frac{\partial\left[rk_i\left(\mathbf{v}, \mathbf{W}\right)\right]}{\partial \mathbf{w}_j} \right] \tag{51}$$

$$= \sum_i p_{\mathbf{v},i} | \mathbf{w} \cdot \left[\xi\left(\mathbf{v}, \mathbf{w}_i\right) h_\lambda\left(rk_i\left(\mathbf{v}, \mathbf{W}\right)\right) \cdot \hat{\delta}_{i,j} \cdot \left(\frac{\partial \xi\left(\mathbf{v}, \mathbf{w}_i\right)}{\partial \mathbf{w}_j} \cdot \sum_l \delta_{il}\left(\mathbf{v}\right) \right) \right] \tag{52}$$

$$- \sum_i p_{\mathbf{v},i} | \mathbf{w} \cdot \left[\xi\left(\mathbf{v}, \mathbf{w}_i\right) h_\lambda\left(rk_i\left(\mathbf{v}, \mathbf{W}\right)\right) \cdot \frac{\partial \xi\left(\mathbf{v}, \mathbf{w}_j\right)}{\partial \mathbf{w}_j} \cdot \delta_{ij}\left(\mathbf{v}\right) \right] \tag{53}$$

$$= p_{\mathbf{v},j} | \mathbf{w} \cdot \left[\xi\left(\mathbf{v}, \mathbf{w}_j\right) h_\lambda\left(rk_j\left(\mathbf{v}, \mathbf{W}\right)\right) \cdot \left(\frac{\partial \xi\left(\mathbf{v}, \mathbf{w}_j\right)}{\partial \mathbf{w}_j} \cdot \delta_{jj}\left(\mathbf{v}\right) \right) \right] \tag{54}$$

$$- p_{\mathbf{v},j} | \mathbf{w} \cdot \left[\xi\left(\mathbf{v}, \mathbf{w}_j\right) h_\lambda\left(rk_j\left(\mathbf{v}, \mathbf{W}\right)\right) \cdot \frac{\partial \xi\left(\mathbf{v}, \mathbf{w}_j\right)}{\partial \mathbf{w}_j} \cdot \delta_{jj}\left(\mathbf{v}\right) \right] \tag{55}$$

$$= 0 \tag{56}$$

which concludes the proof.

Immediate Reward Reinforcement Learning for Clustering and Topology Preserving Mappings

Colin Fyfe and Wesam Barbakh

Applied Computational Intelligence Research Unit,
The University of the West of Scotland, Scotland
{colin.fyfe,wesam.barbakh}@uws.ac.uk

Abstract. We extend a reinforcement learning algorithm which has previously been shown to cluster data. Our extension involves creating an underlying latent space with some pre-defined structure which enables us to create a topology preserving mapping. We investigate different forms of the reward function, all of which are created with the intent of merging local and global information, thus avoiding one of the major difficulties with e.g. K-means which is its convergence to local optima depending on the initial values of its parameters. We also show that the method is quite general and can be used with the recently developed method of stochastic weight reinforcement learning[14].

1 Introduction

There has been a great deal of recent interest in exploratory data analysis mainly because we are automatically acquiring so much data from which we are extracting little information. Such data is typically high-dimensional and high volume, both of which features cause substantial problems. Many of the methods try to project the data to lower dimensional manifolds; in particular, the set of techniques known as exploratory projection pursuit (EPP) [5,4,10] are of interest. These can be thought of as extensions of principal component analysis (PCA) in which the projection sought is one which maximises some projection index so that variance would act as the index for PCA. There have been several 'neural' implementations of EPP [7,9]. However these methods require us to identify what type of structure we are looking for *a priori*: if we are searching for outliers, we use one index, while for clusters, we use an entirely different index. Thus the human operator is very much required when these tools are used for data analysis: indeed, we have often mentioned in our papers that we are specifically using the human's visual pattern matching abilities and the computer's computational abilities optimally in partnership.

An alternative type of exploratory data analysis attempts to cluster the data in some way while the more sophisticated versions of such algorithms attempt to create some (global) ordering of the clusters so that cluster prototypes which are capturing similar data (where again similarity is determined *a priori* by the human user), are shown to do so in some global ordering of the clusters. An early

M. Biehl et al.: (Eds.): Similarity-Based Clustering, LNAI 5400, pp. 35–51, 2009.

and still widely used example of this type of map is Kohonen's Self Organizing Map [12].

We consider the reinforcement learning paradigm interesting as a potential tool for exploratory data analysis: the exploitation-exploration trade-off is exactly what is required in such situations: it precisely matches what a human would do to explore a new data set - investigate, look for patterns *of any identifiable type* and follow partial patterns till they become as clear as possible. This chapter does not fulfil that promise but does investigate a particular form of reinforcement learning as a tool for creating clusters and relating the structure of clusters to one another.

The structure of this chapter is as follows: we first review reinforcement learning and in particular, immediate reward reinforcement learning. We show how this technique can be used to create a topology preserving map by defining latent points' positions at specific locations in an underlying latent space. We then show how the reinforcement learning technique can be used with a number of different reward functions all of which enable clustering to be performed and again, all of which can be used for visualisation if we specify a prior latent space structure. Finally we show how a recent form of immediate reward reinforcement learning can also be used for clustering.

2   Immediate Reward Reinforcement Learning

Reinforcement learning [15] is appropriate for an agent which is actively exploring its environment and also actively exploring what actions are best to take in different situations. Reinforcement learning is so-called because, when an agent performs a beneficial action, it receives some reward which reinforces its tendency to perform that beneficial action again.

There are two main characteristics of reinforcement learning:

1. Trial-and-error search. The agent performs actions appropriate to a given situation without being given instructions as to what actions are best. Only subsequently will the agent learn if the actions taken were beneficial or not.
2. Reward for beneficial actions. This reward may be delayed because the action though leading to a reward may not be (and typically is not) awarded an immediate reward.

Since the agent has a defined goal, as it plays, it will learn that some actions are more beneficial than others in a specific situation. However this raises the exploitation/exploration dilemma: should the agent continue to use a particular action in a specific situation or should it try out a new action in the hope of doing even better. Clearly the agent would prefer to use the best action it knows about for responding to a specific situation but it does not know whether this action is actually optimal unless it has tried every possible action when it is in that situation. This dilemma is sometimes solved by using ϵ-greedy policies which stick with the currently optimal actions with probability 1-ϵ but investigate an alternative action with probability ϵ.

Henceforth we will call the situation presented to the agent, the state of the environment. Note that this state includes not only the passive environment itself but also any changes which may be wrought by other agencies (either other artificial agents or humans) acting upon the environment. This is sometimes described as the environment starts where the direct action of the agent stops i.e. it is everything which the agent cannot directly control. Every state has a value associated with it. This value is a function of the reward which the agent gets from being in that state but also takes into account any future rewards which it may expect to get from its actions in moving from that state to other states which have their own associated rewards. We also create a value function for each action taken in each state.

[17,16] investigated a particular form of reinforcement learning in which reward for an action is immediate which is somewhat different from mainstream reinforcement learning [15,11]. Williams [16] considered a stochastic learning unit in which the probability of any specific output was a parameterised function of its input, $\mathbf{x}$. For the ith unit, this gives

$$P(y_i = \zeta | \mathbf{w}_i, \mathbf{x}) = f(\mathbf{w}_i, \mathbf{x}) \tag{1}$$

where, for example,

$$f(\mathbf{w}_i, \mathbf{x}) = \frac{1}{1 + \exp(- \| \mathbf{w}_i - \mathbf{x} \|^2)} \tag{2}$$

Williams [16] considers the learning rule

$$\Delta w_{ij} = \alpha_{ij}(r_{i,\zeta} - b_{ij}) \frac{\partial \ln P(y_i = \zeta | \mathbf{w}_i, \mathbf{x})}{\partial w_{ij}} \tag{3}$$

where α_{ij} is the learning rate, $r_{i,\zeta}$ is the reward for the unit outputting ζ and b_{ij} is a reinforcement baseline which in the following we will take as the reinforcement comparison, $b_{ij} = \bar{r} = \frac{1}{K} \sum r_{i,\zeta}$ where K is the number of times this unit has output ζ. ([16], Theorem 1) shows that the above learning rule causes weight changes which maximise the expected reward.

[16] gave the example of a Bernoulli unit in which $P(y_i = 1) = p_i$ and so $P(y_i = 0) = 1 - p_i$. Therefore

$$\frac{\partial \ln P(y_i)}{\partial p_i} = \begin{cases} -\frac{1}{1-p_i} & \text{if } y_i = 0 \\ \frac{1}{p_i} & \text{if } y_i = 1 \end{cases} = \frac{y_i - p_i}{p_i(1 - p_i)} \tag{4}$$

[13] applies the Bernoulli model to (unsupervised) clustering with

$$p_i = 2(1 - f(\mathbf{w}_i, \mathbf{x})) = 2(1 - \frac{1}{1 + \exp(- \| \mathbf{w}_i - \mathbf{x} \|^2)}) \tag{5}$$

The environment identifies the p_{i^*} which is maximum over all output units and y_{i^*} is then drawn from this distribution. Rewards are given such that

$$r_i = \begin{cases} 1 & \text{if } i = i^* \text{ and } y_i = 1 \\ -1 & \text{if } i = i^* \text{ and } y_i = 0 \\ 0 & \text{if } i \neq i^* \end{cases} \tag{6}$$

This is used in the update rule

$$\Delta w_{ij} = \alpha r_i (y_i - p_i)(x_j - w_{ij}) \tag{7}$$
$$= \alpha |y_i - p_i|(x_j - w_{ij}) \text{ for } i = i^* \tag{8}$$

which is shown to perform clustering of the data set.

2.1   Topology Preserving Maps

Topology preserving maps [12,6] can also be thought of as clustering techniques but clustering techniques which attempt to capture some structure in a data set by creating a visible relationship between nearby clusters. Most research has been into methods which modify an adaptive clustering rule in a way that preserves neighbouring relationships. However an alternative [3,8] is to create a latent space of points $\mathbf{t}_1, ..., \mathbf{t}_K$ which *a priori* have some structure such as lying equidistantly placed on a line or at the corners of a grid. These are then mapped non-linearly to the data space; the nonlinearity is essential since the data clusters need not be constrained to lie on a line. Thus we [3,8] map the latent points through a set of basis functions, typically squared exponentials centered in latent space, and then map the output of the basis functions through a set of weights to points, $\mathbf{m}_1, ..., \mathbf{m}_K$, in data space. Let there be M basis functions, $\phi_j(), j = 1, ..., M$, with centres, μ_j in latent space; therefore, using $\mathbf{w}_j$ as the weight from the jth basis function to data space,

$$\mathbf{m}_k = \sum_{j=1}^{M} \mathbf{w}_j \phi_j(\mathbf{t}_k) = \sum_{j=1}^{M} \mathbf{w}_j \exp(-\beta \parallel \mu_j - \mathbf{t}_k \parallel^2) \tag{9}$$

Since $\frac{\partial \mathbf{m}_k}{\partial w_{ij}} = \phi_j(\mathbf{t}_k)$, with the Bernoulli learner (8) this gives the learning rule

$$\Delta w_{dj} = \alpha |y_{i^*} - p_{i^*}|(x_d - \mathbf{w}_{jd}\phi_j(\mathbf{t}_{i^*}))\phi_j(\mathbf{t}_{i^*}) \tag{10}$$

In Figure 1, we illustrate a one dimensional latent space mapped to a two dimensional data set, in which $\mathbf{x}_1$ evenly divides the interval $[-\pi/2, \pi/2]$ and $\mathbf{x}_2 = \mathbf{x}_1 + 1.25 \sin(\mathbf{x}_1) + \mu$ where μ is noise from a uniform distribtion in $[0,0.5]$; we see that the one dimensional nature of the manifold has been identified and the latent points projections into data space have maintained their ordering. The left diagram shows the probabilities which each latent point takes for each data point (arranged in ascending order along the manifold); we see a sensible transition of probabilities as we cross the latent space.

2.2   The Gaussian Learner

One advantage that this representation has is that it is very simple to change the base learner. One type of learner which is evidently appropriate because of the development of the model [3,8] is the Gaussian. Again let i^* define the winning (closest) prototype or centre. We draw a sample from the i^* distribution so

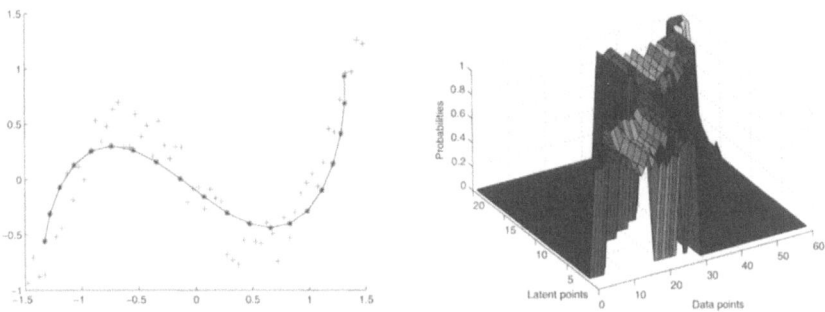

Fig. 1. Left: the data set and converged centres. Right: the probabilities each centre accords to each data point.

that $y \sim N(\mathbf{m}_{i^*}, \beta_{i^*}^2)$, the Gaussian distribution with mean $\mathbf{m}_{i^*}$ and variance $\beta_{i^*}^2$. Each learner now has two parameters to learn, its mean and variance. The learning rules can be derived [16] as

$$\Delta\mathbf{m} = \alpha_m (r - \bar{r}) \frac{\| \mathbf{y} - \mathbf{m} \|}{\beta^2} \tag{11}$$

$$\Delta\beta = \alpha_\beta (r - \bar{r}) \frac{\| \mathbf{y} - \mathbf{m} \|^2 - \beta^2}{\beta^3} \tag{12}$$

where we calculate the reward, $r = \exp(- \| \mathbf{x} - \mathbf{y} \|^2)$, i.e. the closer the data sample is to the distribution sample, the greater the reward. Of course, for visualisation we are updating the weights again rather than the prototypes so (11) becomes

$$\Delta w_{dj} = \alpha_m (r_{y_{i^*}} - \overline{r_{i^*}}) \frac{\| \mathbf{y}_{i^*,d} - \mathbf{m}_{i^*,d} \| \phi_{i^*,d}}{\beta_{i^*}^2} \tag{13}$$

Note the effects of the learning rules on the Gaussian parameters. If a value y is chosen which leads to a better reward than has typically been achieved in the past, the change in the mean is towards y; if the reward is less than the previous average, change is away from y. Also, if the reward is greater than the previous average, the variance will decrease if $\| \mathbf{y} - \mathbf{m} \|^2 < \beta^2$ i.e. narrowing the search while it will increase if $\| \mathbf{y} - \mathbf{m} \|^2 > \beta^2$, thus widening the search volume. We have found that it is best to update the variances more slowly than the means.

We have found that the parameters of this model requires more careful tuning than those of the Bernoulli learner. Results from an example simulation are shown in Figure 2. We see that the variances at the extrema remain somewhat larger than those in the centre of the data.

2.3   Visualisation

We conclude this section with an example simulation showing the method used for visualisation purposes. The data set is composed of 72 samples of algae

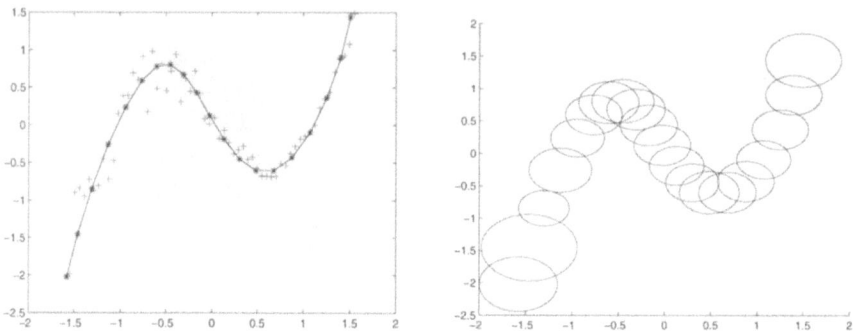

Fig. 2. Left: The data ('+'s) and the centres ('*'s). Right: each circle represents one standard deviation centered on the centres shown in the first figure.

for each of which we have 18 floating point values denoting their pigments at different frequencies; the data has been classified into 9 different classes though this information is not given to clustering method. We use the Bernoulli learner with an underlying two dimensional latent space and train using the above rules. Results are shown in Figure 3. We see a reasonable separation of the classes comparable to results we have achieved with other methods. To project into the latent space in the left diagram, we have used the positions of the latent points in their two dimensional space weighted by the probabilities they gave to each data point. This gives a finer resolution mapping than simply plotting each point at the latent space point which gave it highest probability.

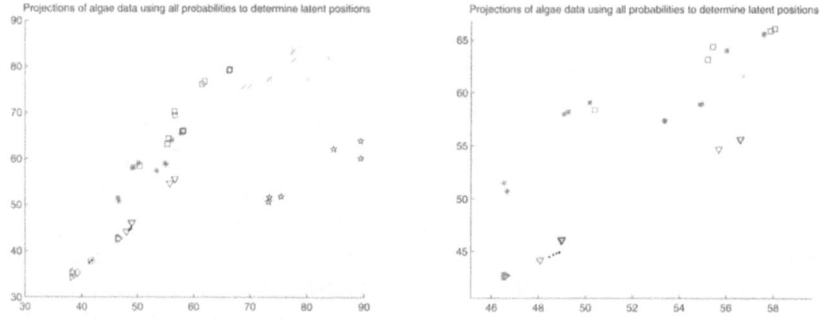

Fig. 3. Left: using all the probabilities to map the algae to latent space. Right: zooming in to the centre of the latent projections.

3   Alternative Reward Functions

In this section, we investigate using reinforcement learning with reward functions which are related to our previous clustering methods in that the reward functions

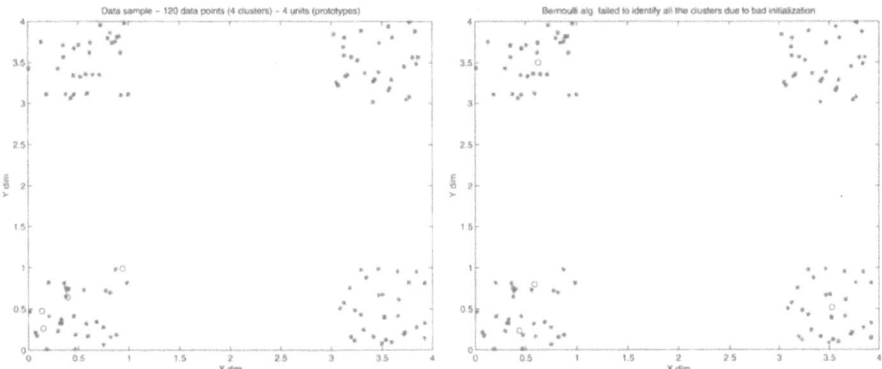

Fig. 4. Left: artificial data set is shown as 6 clusters of '*'s, and 6 prototypes of 'o's. Right: Bernoulli algorithm failed to identify all the clusters successfully.

also allow the reinforcement learning algorithms to overcome the disadvantages of a poor initialization and achieve the globally optimal clustering.

We applied the Bernoulli algorithm [13] to the artificial data set shown in Figure 4, left, but the Bernoulli algorithm failed to identify all the clusters successfully as shown in Figure 4, right.

The Bernoulli algorithm also suffers from a poor initialization and shows sensitivity to the prototypes' initialization. This results in dead prototypes and convergence to a local optimum. The main reason for these problems is that we update in this algorithm the winner prototypes only, not all of them. Poor initialization may allow some prototypes to learn while others don't respond or learn at all. Thus we illustrate three different reward functions [2] that overcome the problem of poor initialization and the convergence to the local optimum. These new reward functions allow all the nodes to learn, not only the winners.

Note that while it is unlikely that we would initialize all prototypes to a single cluster in any real data set, it is quite possible that some clusters may be omitted in the initial seeding of prototypes. Such clusters may never be found using the Bernoulli algorithm (or indeed standard K-means).

3.1 New Algorithm RL1

The RL1 algorithm has the following reward function which is motivated by attempting to incorporate both local and global information in the clustering process:

$$r_i = \begin{cases} \frac{\|\mathbf{x}-\mathbf{m}_{k*}\|^3}{\|\mathbf{x}-\mathbf{m}_i\|^3} & \text{if } y_i = 1 \\ -\frac{\|\mathbf{x}-\mathbf{m}_{k*}\|^3}{\|\mathbf{x}-\mathbf{m}_i\|^3} & \text{if } y_i = 0 \end{cases} \tag{14}$$

where

$$k* = \arg\min_{k=1}^{K}(\| \mathbf{x} - \mathbf{m}_k \|)$$

This new reward function has the following advantages:

1. We are rewarding all prototypes (nodes), not only the winners and thus all prototypes will learn to find the clusters even if they are initialized badly.
2. This reward function allows the prototypes to respond differently to each other, and each prototype, before moving to any new locations, responds to all the other prototypes' positions, and hence it is possible for it to identify the free clusters that are not recognized by the other prototypes.
3. This reward function gives the highest value, 1, for the highest similarity between the data point and the node (prototype).

Simulations. Figure 5 shows the result after applying the RL1 algorithm to the artificial data set, but with very poor prototypes' initialization.

Figure 5, left, shows the prototypes after many iterations but before convergence; in this Figure, we can see one prototype still distant from the data points while others have spread into the data; this distant prototype still has the ability to learn even if it is very far from data, and this is an advantage for this algorithm over the previous algorithms.

3.2   Second Algorithm RL2

The RL2 algorithm has the following reward function which is motivated by inverse exponential K-means algorithm [1]:

$$r_i = \begin{cases} 1 & \text{if } i = k^* \text{ and } y_i = 1 \\ \\ \frac{1-\exp(-\beta\|\mathbf{x}-\mathbf{m}_{k*}\|^3)}{\|\mathbf{x}-\mathbf{m}_i\|^3} & \text{if } i \neq k^* \text{ and } y_i = 1 \\ \\ -1 & \text{if } i = k^* \text{ and } y_i = 0 \\ \\ \frac{\exp(-\beta\|\mathbf{x}-\mathbf{m}_{k*}\|^3)-1}{\|\mathbf{x}-\mathbf{m}_i\|^3} & \text{if } i \neq k^* \text{ and } y_i = 0 \end{cases} \qquad (15)$$

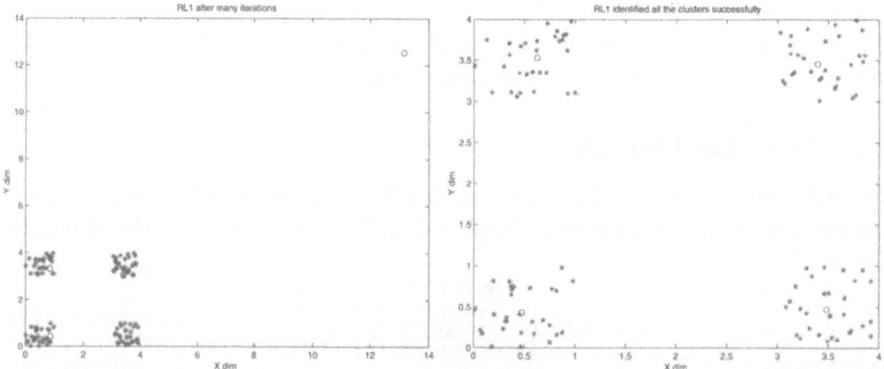

Fig. 5. Left: RL1 result after many iterations but before convergence. Right: RL1 result after convergence.

where again $k* = \arg\min_j \| \mathbf{x} - \mathbf{m}_j \|$.

The reward function (15) has values ranged between 0 and 1. We need to update the closest prototype (or most similar one) by giving it directly the maximum possible reward value, which equals 1, to allow it to learn more than others and also to avoid any division by zero, which may happen using the second equation in (15). The second equation in (15) is used for all the other prototypes. Prototypes closer (or more similar) to the input data sample will learn more than others by taking a higher reward value, thus all prototypes learn appropriately.

3.3   Third Algorithm, RL3

The RL3 algorithm has the following reward function which is motivated by K-Harmonic means[19,18]:

$$
r_i = \begin{cases} \dfrac{1}{\|\mathbf{x}-\mathbf{m}_i\|^4\{\sum_{l=1}^{K}\frac{1}{\|\mathbf{x}-\mathbf{m}_l\|^2}\}^2} & \text{if } y_i = 1 \\[4mm] \dfrac{-1}{\|\mathbf{x}-\mathbf{m}_i\|^4\{\sum_{l=1}^{K}\frac{1}{\|\mathbf{x}-\mathbf{m}_l\|^2}\}^2} & \text{if } y_i = 0 \end{cases} \tag{16}
$$

The reward function in (16) has similar principles to the previous new reward functions. It has values ranged between 0 and 1. All the prototypes can learn in an effective way. The prototype that is more similar to the input data sample gets a higher reward value. In implementation, to avoid any division by zero we can rewrite (16) as follows:

$$
r_i = \begin{cases} \dfrac{\frac{\|\mathbf{x}-\mathbf{m}_{k*}\|^4}{\|\mathbf{x}-\mathbf{m}_i\|^4}}{\left\{1+\sum_{l\neq k*}^{K}\frac{\|\mathbf{x}-\mathbf{m}_{k*}\|^2}{\|\mathbf{x}-\mathbf{m}_l\|^2}\right\}^2} & \text{if } y_i = 1 \\[6mm] \dfrac{-\frac{\|\mathbf{x}-\mathbf{m}_{k*}\|^4}{\|\mathbf{x}-\mathbf{m}_i\|^4}}{\left\{1+\sum_{l\neq k*}^{K}\frac{\|\mathbf{x}-\mathbf{m}_{k*}\|^2}{\|\mathbf{x}-\mathbf{m}_l\|^2}\right\}^2} & \text{if } y_i = 0 \end{cases} \tag{17}
$$

where $k* = \arg\min_j \| \mathbf{x} - \mathbf{m}_j \|$.

Notice: $\frac{\|\mathbf{x}-\mathbf{m}_{k*}\|^4}{\|\mathbf{x}-\mathbf{m}_{k*}\|^4}$ is always set to 1.

Simulations. Figure 6 shows the results after applying the Bernoulli algorithm, top right, RL2, bottom left, and RL3, bottom right, to the artificial data set shown in Figure 6, top left. RL2 and RL3 succeed in identifying the clusters successfully while the Bernoulli model failed.

To show how these algorithms behave with dead prototypes, we have Figure 7, left, which contains some. Figure 7, right, shows the result after applying the Bernoulli algorithm to the same artificial data set as Figure 6, top left, but with very poor prototypes' initialization as shown in Figure 7, left. The Bernoulli algorithm gave bad results and there are 7 dead prototypes which don't learn. Figure 8 shows the result after applying the RL2 algorithm to the same artificial data set. From Figure 8, top and bottom left, we can see some prototypes still distant from the data points while others spread into data. These distant prototypes still have

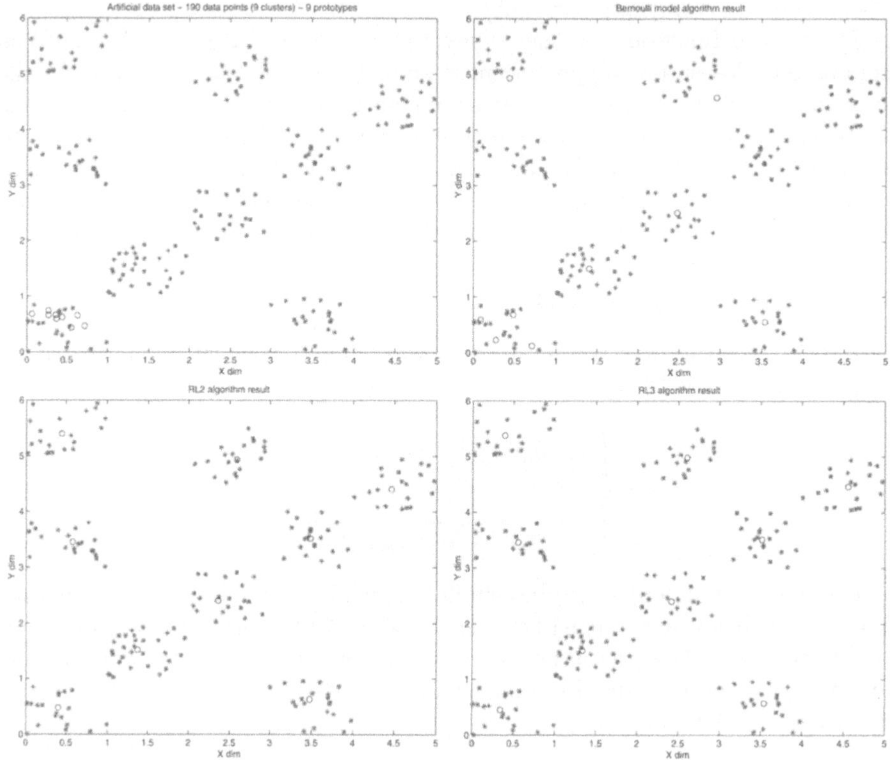

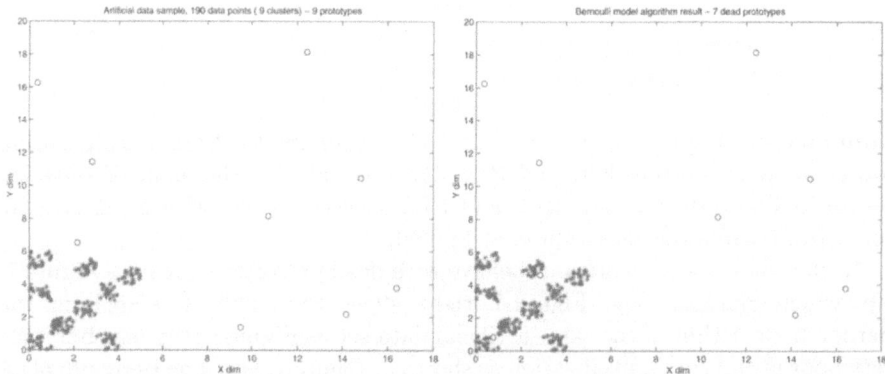

Fig. 6. Top left: artificial data set with poor prototypes' initialization. Top right: the Bernoulli algorithm result. Bottom left: RL2 algorithm result. Bottom right: RL3 algorithm result.

Fig. 7. Left: artificial data set with poor prototypes' initialization. Right: Bernoulli algorithm result, failed to identify clusters - 7 dead prototypes.

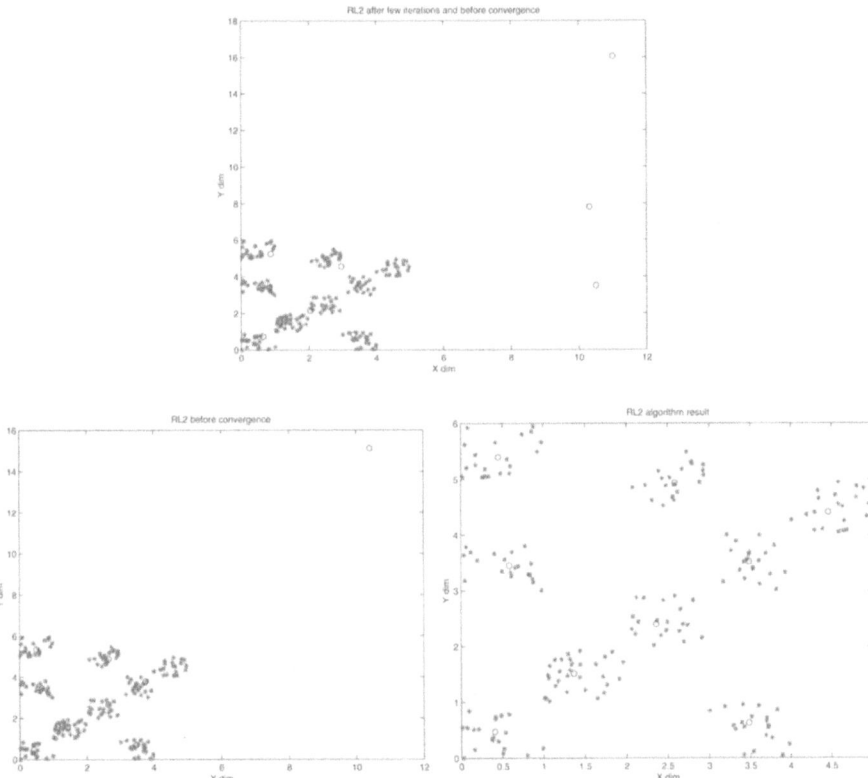

Fig. 8. RL2 algorithm results. Top: result after 10 iterations and before convergence. Bottom left: result after 30 iterations and before convergence. Bottom right: RL2 algorithm result after 70 iterations.

the ability to learn and identify clusters, as shown in Figure 8, bottom right, even if they are far from data. This activation of dead prototypes is an advantage of these new algorithms.

3.4 Topology Preserving Mapping

Again we use the same structure as the GTM to visualize data, but this time with RL1 and RL2 for the learning process to construct RL1ToM and RL2ToM.

Artificial data set. We create a simulation with 20 latent points deemed to be equally spaced in a one dimensional latent space, passed through 5 Gaussian basis functions and then mapped to the data space by the linear mapping W which is the only parameter we adjust. We illustrate with a similar artificial data set to the previous section. Final result from the RL1ToM is shown in Figure 9.

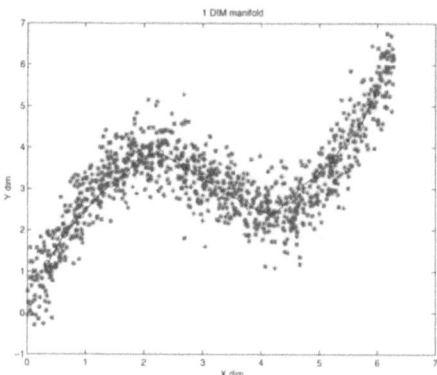

Fig. 9. The resulting prototypes' positions after applying RL1ToM. Prototypes are shown as 'o's.

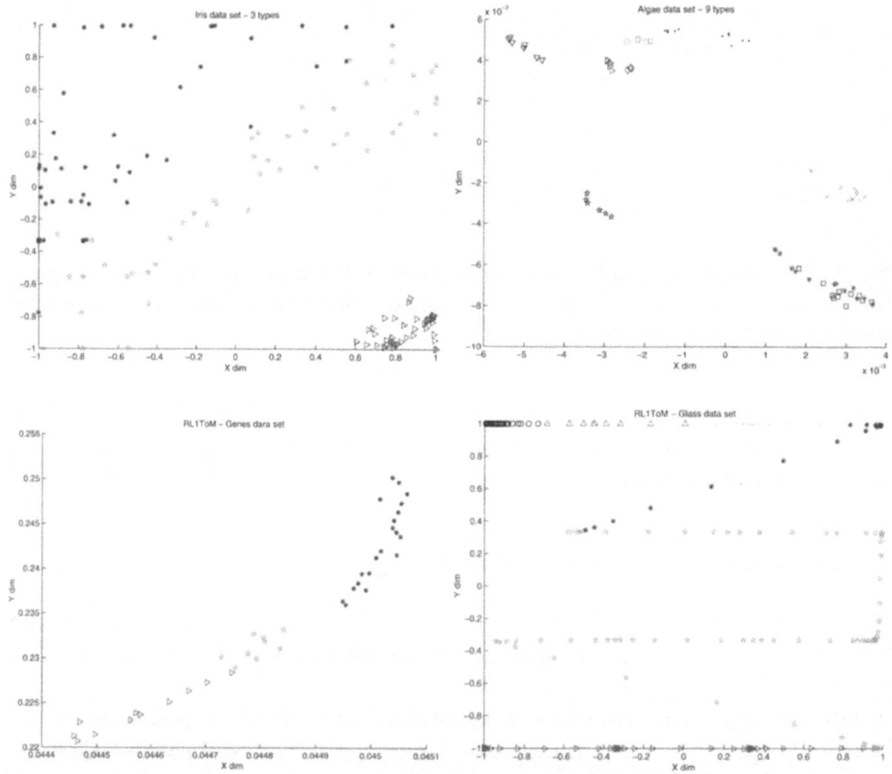

Fig. 10. Visualisation using the RL1ToM on 4 real data sets

Real data sets. We show in Figure 10 and Figure 11 the projections of the real data sets, iris, algae, genes and glass, onto a two dimensional grid of latent points (10 x 10) using RL1ToM and RL2ToM, respectively. The results are comparable with others we have with these data sets from a variety of different algorithms. The RL2ToM algorithm gives the best result, similar to those of IKoToM, for the algae data set. It succeeds in visualizing all the 9 clusters successfully, while most of the previous methods failed to identify all of them.

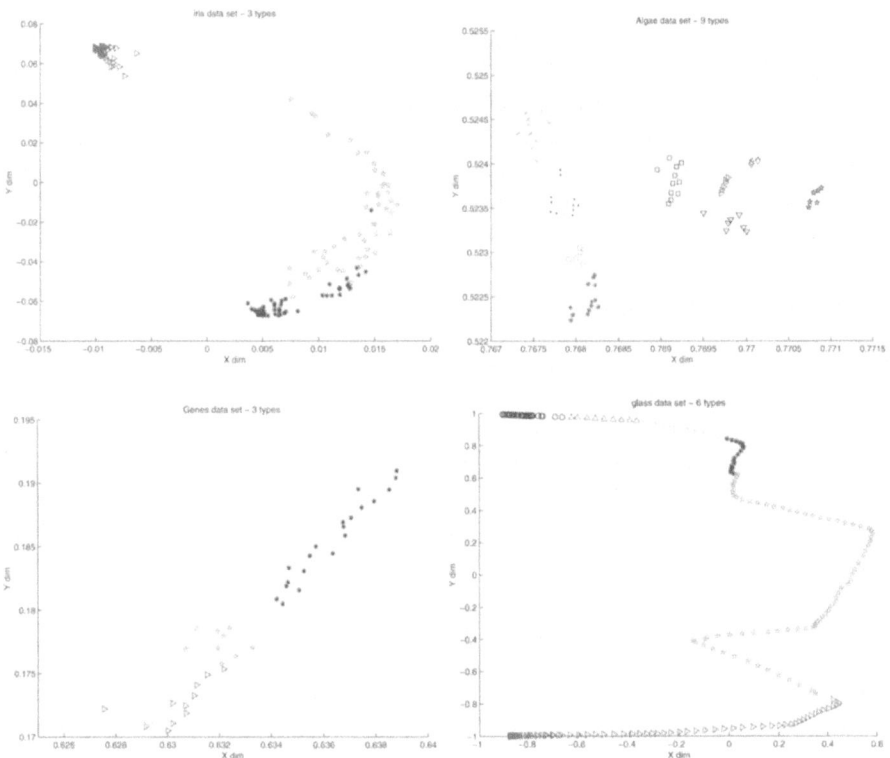

Fig. 11. Visualisation using the RL2ToM on 4 real data sets

3.5   Discussion

We have discussed one shortcoming of the K-means algorithm: its sensitivity to poor initialization which leads it to converge to a local rather than global optimum. We have shown how different performance functions lead to algorithms which incorporate the correct mixture of local and global knowledge to allow prototypes to optimally cluster a data set.

We have extended these algorithms by using them with an underlying latent space which enables topology preserving mappings to be developed. We have illustrated these mappings on a variety of data sets and shown how they may be used to visualise these data sets.

We have shown how reinforcement learning of cluster prototypes can be performed robustly by altering the reward function associated with finding the clusters. We have illustrated three different reward functions which clearly have a family resemblance. Most importantly all three overcome the disadvantages of poor initialization in that they do not succumb to local minima as the existing Bernoulli algorithm does.

It is worth stating that the original Bernoulli algorithm [13] seems to somewhat cheat in its formulation of the clustering method: we already know that the winning node is the closest and hence that it should be emitting 1 and so every update moves the node's parameters closer to the data point for which it was deemed to be closest. However the algorithms herein sample each and every node using the distribution corresponding to its current parameters i.e. no prior assumptions about the winning node are made except in the formulation of the reward which has the winning distance as numerator of the reward.

We have also illustrated how a topology preserving mapping can be created by using these algorithms with an underlying fixed latent space.

4   Stochastic Weights

An alternative view of supervised networks with reinforcement learning is given in [14]: the networks are composed of deterministic nodes but stochastic synapses. There are two variants of this method: in the first, [14] uses a set of weights, w_{ij}, each of which is drawn from a Gaussian distribution $N(\mu_{ij}, \sigma_{ij}^2)$ with mean μ_{ij} and variance, σ_{ij}^2. Then the output of this deterministic neuron is also a Gaussian random variable with mean $\mu_{y_i} = \sum_j \mu_{ij} x_j$ and variance $\sigma_{y_i}^2 = \sum_j \sigma_{ij}^2 x_j^2$. With this model, [14] shows that the learning rule

$$\Delta \mu_{ij} = \eta r \frac{(y_i - \mu_{y_i}) x_j}{\sigma_{y_i}^2} \qquad (18)$$

where η is the learning rate, follows the gradient of r, the reward given to the neuron. i.e. we are performing gradient ascent on the reward function by changing the parameters using (18). Often a simplified form is used by incorporating $\sigma_{y_i}^2$ into the learning rate:

$$\Delta \mu_{ij} = \eta r (y_i - \mu_{y_i}) x_j \qquad (19)$$

[14] also investigate an additional term which they say enables the rule to escape from local optima:

$$\Delta \mu_{ij} = \eta [r(y_i - \mu_{y_i}) + \lambda(1 - r)(-y_i - \mu_{y_i})] x_j \qquad (20)$$

[14] denotes (19) and (20) by rules A1 and A2 respectively.

In this first view, the neuron's output is a function of two independent sets of variables, the inputs and the weights. In the second view, the input is deemed to be fixed and only the synapses are thought of as stochastic. Then $p(w_{ij}) = Z \exp(-\frac{(w_{ij} - \mu_{ij})^2}{2\sigma_{ij}^2})$ which leads to a second set of learning rules:

$$\Delta \mu_{ij} = \eta r (w_{ij} - \mu_{ij}) \qquad (21)$$

which again may include the local optima avoidance term to give

$$\Delta\mu_{ij} = \eta[r(w_{ij} - \mu_{ij}) + \lambda(1 - r)(-w_{ij} - \mu_{ij})] \tag{22}$$

[14] denotes (21) and (22) by rules B1 and B2 respectively. These rules are tested on a number of standard problems in [14].

4.1   Simulation

Thus we [3,8] map the latent points through a set of basis functions, typically squared exponentials centered in latent space, and then map the output of the basis functions through a set of weights to points, $\mathbf{m}_1, ..., \mathbf{m}_K$, in data space. Let there be M basis functions, $\phi_j(), j = 1, ..., M$, with centres, μ_j in latent space; therefore, using $\mathbf{w}_j$ as the weight from the jth basis function to data space,

$$\mathbf{m}_k = \sum_{j=1}^{M} \mathbf{w}_j \phi_j(\mathbf{t}_k) = \sum_{j=1}^{M} \mathbf{w}_j \exp(-\beta \parallel \mu_j - \mathbf{t}_k \parallel^2), \forall k \in \{1, ..., K\} \tag{23}$$

Since $\frac{\partial \mathbf{m}_k}{\partial w_{ij}} = \phi_j(\mathbf{t}_k)$, with the Gaussian model A1 above

$$\Delta\mu_{i*} = \eta r(\mathbf{w} - \mu_{y_{i*}})\phi(i^*) \tag{24}$$

where i^* identifies the agent with closest centre, $\mathbf{m}_{i*}$ to the current data point, $\mathbf{x}$. $\mathbf{w}$ is a sample from $\mu_{y_{i*}}$ and the reward is $r = \exp(- \parallel \mathbf{x} - \mathbf{w} \parallel)$.

We illustrate (Figure 12) a one dimensional latent space mapped to a two dimensional data set, $\mathbf{x} = (\mathbf{x}_1, \mathbf{x}_2)$, in which $\mathbf{x}_1$ evenly divides the interval $[-\pi/2, \pi/2]$ and $\mathbf{x}_2 = \mathbf{x}_1 + 1.25\sin(\mathbf{x}_1) + \mu$ where μ is noise from a uniform

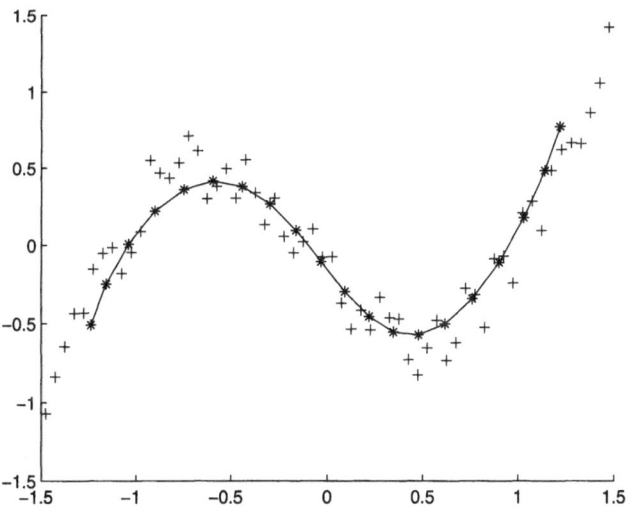

Fig. 12. The data are shown by '+'s; the latent points' projections are shown by '*'s

distribtion in [0,0.5]; we see that the one dimensional nature of the manifold has been identified and the latent points' projections into data space have maintained their ordering. This simulation used rule A2 and we can see one of the effects of the local minimum avoidance term which is a drawing in of the map to the centre of the data. However there are local optima with these maps (the Kohonen SOM often exhibits twists across the data) and so in practice it is best to use this term but with a decay of the λ parameter during the course of the simulation till rule A1 is reached.

5   Conclusion

We have taken an existing method [13] for creating clusters using immediate reinforcement learning and applied it with an underlying latent space. Since the points in the latent space are static, this forces the clustering operation to maintain some form of neighbourhood relations. We have illustrated two base learners, the Bernoulli learner and the Gaussian learner. Finally we have shown an example of visualisation on an 18 dimensional data set. We have illustrated clustering and creating such maps with a variety of reward functions, all designed to incorporate global and local information in the clustering process. Finally we showed that the recently developed stochastic weight method [14] also can be used for clustering and topology preserving mappings.

References

1. Barbakh, W.: Local versus Global Interactions in Clustering Algorithms. Ph.D thesis, School of Computing, University of the West of Scotland (2008)
2. Barbakh, W., Fyfe, C.: Clustering with reinforcement learning. In: Yin, H., Tino, P., Corchado, E., Byrne, W., Yao, X. (eds.) IDEAL 2007. LNCS, vol. 4881, pp. 507–516. Springer, Heidelberg (2007)
3. Bishop, C.M., Svensen, M., Williams, C.K.I.: Gtm: The generative topographic mapping. Neural Computation (1997)
4. Friedman, J.H.: Exploratory projection pursuit. Journal of the American Statistical Association 82(397), 249–266 (1987)
5. Friedman, J.H., Tukey, J.W.: A projection pursuit algorithm for exploratory data analysis. IEEE Transactions on Computers c-23(9), 881–889 (1974)
6. Fyfe, C.: A scale invariant feature map. Network: Computation in Neural Systems 7, 269–275 (1996)
7. Fyfe, C.: A comparative study of two neural methods of exploratory projection pursuit. Neural Networks 10(2), 257–262 (1997)
8. Fyfe, C.: Two topographic maps for data visualization. Data Mining and Knowledge Discovery 14, 207–224 (2007)
9. Intrator, N.: Feature extraction using an unsupervised neural network. Neural Computation 4(1), 98–107 (1992)
10. Jones, M.C., Sibson, R.: What is projection pursuit. Journal of The Royal Statistical Society, 1–37 (1987)
11. Kaelbling, L.P., Littman, M.L., Moore, A.W.: Reinforcement learning: A survey. Journal of Artificial Intelligence Research 4, 237–285 (1996)

12. Kohonen, T.: Self-Organising Maps. Springer, Heidelberg (1995)
13. Likas, A.: A reinforcement learning approach to on-line clustering. Neural Computation (2000)
14. Ma, X., Likharev, K.K.: Global reinforcement learning in neural networks with stochastic synapses. IEEE Transactions on Neural Networks 18(2), 573–577 (2007)
15. Sutton, R.S., Barto, A.G.: Reinforcement Learning: an Introduction. MIT Press, Cambridge (1998)
16. Williams, R.: Simple statistical gradient-following algorithms for connectionist reinforcement learning. Machine Learning 8, 229–256 (1992)
17. Williams, R.J., Pong, J.: Function optimization using connectionist reinforcement learning networks. Connection Science 3, 241–268 (1991)
18. Zhang, B.: Generalized k-harmonic means – boosting in unsupervised learning. Technical report, HP Laboratories, Palo Alto (October 2000)
19. Zhang, B., Hsu, M., Dayal, U.: K-harmonic means - a data clustering algorithm. Technical report, HP Laboratories, Palo Alto (October 1999)

Advances in Feature Selection with Mutual Information

Michel Verleysen[1], Fabrice Rossi[2], and Damien François[1]

[1] Université catholique de Louvain, Machine Learning Group,
[2] INRIA Rocquencourt, Domaine de Voluceau, Rocquencourt, B.P. 105,
78153 Le Chesnay Cedex, France
Michel.Verleysen@uclouvain.be, Fabrice.Rossi@inria.fr,
Damien.Francois@uclouvain.be

Abstract. The selection of features that are relevant for a prediction or classification problem is an important problem in many domains involving high-dimensional data. Selecting features helps fighting the curse of dimensionality, improving the performances of prediction or classification methods, and interpreting the application. In a nonlinear context, the mutual information is widely used as relevance criterion for features and sets of features. Nevertheless, it suffers from at least three major limitations: mutual information estimators depend on smoothing parameters, there is no theoretically justified stopping criterion in the feature selection greedy procedure, and the estimation itself suffers from the curse of dimensionality. This chapter shows how to deal with these problems. The two first ones are addressed by using resampling techniques that provide a statistical basis to select the estimator parameters and to stop the search procedure. The third one is addressed by modifying the mutual information criterion into a measure of how features are complementary (and not only informative) for the problem at hand.

1 Introduction

High-dimensional data are nowadays found in many applications areas: image and signal processing, chemometrics, biological and medical data analysis, and many others. The availability of low cost sensors and other ways to measure information, and the increased capacity and lower cost of storage equipments, facilitate the simultaneous measurement of many features, the idea being that adding features can only increase the information at disposal for further analysis.

The problem is that high-dimensional data are in general more difficult to analyse. Standard data analysis tools either fail when applied to high-dimensional data, or provide meaningless results. Difficulties related to handling high-dimensional data are usually gathered under the *curse of dimensionality* terms, which gather many phenomena usually having counter-intuitive mathematical or geometrical interpretation. The curse of dimensionality already concerns simple phenomena, like colinearity. In many real-world high-dimensional

M. Biehl et al.: (Eds.): Similarity-Based Clustering, LNAI 5400, pp. 52–69, 2009.

problems, some features are highly correlated. But if the number of features exceeds the number of measured data, even a simple linear model will lead to an undetermined problem (more parameters to fit than equations). Other difficulties related to the curse of dimensionality arise in more common situations, when the dimension of the data space is high even if many data are available for fitting or learning. For example, data analysis tools which use Euclidean distances between data or representatives, or any kind of Minkowski or fractional distance (i.e. most tools) suffer from the fact that distances concentrate in high-dimensional spaces (distances between two random close points and between two random far ones tend to converge to the same value, in average).

Facing these difficulties, data analysis tools must address two ways to counteract them. One is to develop tools that are able to model high-dimensional data with a number of (effective) parameters which is lower than the dimension of the space. As an example, Support-Vector Machines enter into this category. The other way is to decrease in some way the dimension of the data space, without significant loss of information. The two ways are complementary, as the first one addresses the algorithms while the second preprocesses the data themselves. Two possibilities also exist to reduce the dimensionality of the data space: features (dimensions) can be selected, or combined. Feature combination means to *project* data, either linearly (Principal Component Analysis, Linear Discriminant Analysis, etc.) or nonlinearly. Selecting features, i.e. keeping some of the original features as such, and discarding others, is a priori less powerful than projection (it is a particular case). However, it has a number of advantages, mainly when interpretation is sought. Indeed after selection the resulting features are among the original ones, which allows the data analyst to interact with the application provider. For example, discarding features may help avoiding to collect useless (possibly costly) features in a further measument campaign. Obtaining relevances for the original features may also help the application specialist to interpret the data analysis results, etc. Another reason to prefer selection to projection in some circumstances, is when the dimension of the data is really high, and the relations between features known or identified to be strongly non-linear. In this case indeed linear projection tools cannot be used; and while nonlinear dimensionality reduction is nowadays widely used for data visualization, its use in quantitative data preprocessing remains limited because of the lack of commonly accepted standard method, the need for expertise to use most existing tools and the computational cost of some of the methods.

This chapter deals with feature selection, based on mutual information between features. The following of this chapter is organized as follows. Section 2 introduces the problem of feature selection and the main ingredients of a selection procedure. Section 3 details the Mutual Information relevance criterion, and the difficulties related to its estimation. Section 4 shows how to solve these issues, in particular how to choose the smoothing parameter in the Mutual Information estimator, how to stop the greedy search procedure, and how to extend the mutual information concept by using nearest neighbor ranks when the dimension of the search space increases.

2   The Two Ingredients of Feature Selection

Feature selection aims at reducing the dimensionality of data. It consists in selecting *relevant* variables (or features) among the set of original ones. The relevance has to be measured in an objective way, through an appropriate and well-defined criterion. However, defining a criterion does not solve the feature selection problem. As the number of initial features is usually large, it is computationally impossible to test all possible subsets of them, even if the criterion to measure the relevance is simple to evaluate. In addition to the definition of the criterion, there is thus a need to define a search procedure among all possible subsets. The relevance criterion and the greedy search procedure are the two basic ingredients of feature selection.

Note that in some situations, feature selection does not aim only at selecting features among the original ones. In some cases indeed, potentially relevant features are not known in advance, and must be extracted or created from the raw data. Think for example of data being curves, as in spectroscopy, in hysteresis curve analysis, or more generally in the processing of functions. In this case the dimension of the data is infinite, and a first choice must consist in extracting a finite number of original features. Curve sampling may be an answer to this question, but other features, as integrals, area under curve, derivatives, etc. may give appropriate information for the problem too. It may thus reveal interesting to first extract a large number of features in a more or less blind way from the original data, and then to use feature selection to select those that are most relevant, in an objective way.

In addition to choosing a relevance criterion and a greedy procedure, a number of other issues have to be addressed. First, one has to define on which features to apply the criterion. For example, if the criterion is correlation, is it better to keep features that are highly correlated to the output (and to drop the other ones), or to drop features that are highly correlated between them (and to keep uncorrelated ones)? Both ideas are reasonable, and will lead to different selections.

Another key issue is simply whether to use a criterion or not. If the goal of feature selection is to use the reduced feature set as input to a prediction or classification model, why not to use the model itself as a criterion? In other words, why not fitting a model on each possible subset (resulting from the greedy search), instead of using a criterion that will probably result in measuring relevance in a different way as the model would do? Using the model is usually referred to as a *wrapper* approach, while using an alternative criterion is a *filter* approach. In theory, there is nothing better than using the model itself, as the final goal is model performances. However, the wrapper way may have two drawbacks: first it could be computationally too intensive for example when using nonlinear neural networks or machine learning tools that require tedious learning. Secondly, when the stochastic nature of the tools makes that their results vary according to initial conditions or other parameters, the results may not be unique, which results in a noisy estimation of the relevance and the need for further simulations to reduce this noise. The main goal of criteria in filters is then to *facilitate* the measure of feature relevance, rather than to provide a unique and unquestionable way of evaluation. This must be kept in mind when designing both the relevance criterion and the greedy procedure: both will act as compromises between adequateness (with the final goal of model performances) and the computational complexity.

Most of the above issues are extensively discussed in the large scientific literature about feature selection. One issue which is much less discussed is how to evaluate the criterion. An efficient criterion must measure any kind of relation between features, not only linear relations. Such a nonlinear criterion is however a (simplified) data model by itself, and requires to fix some design parameters. How to fix these parameters is an important question too, as an inappropriate choice may lead to wrong relevance estimations.

This chapter mainly deals with the last question, i.e. how to estimate in practice, and efficiently, the relevance criterion. Choices that are made concerning the criterion itself and the greedy procedure are as follows.

As the relevance criterion must be able to evaluate any relation between features, and not only linear relations, the correlation is not appropriate. A nonlinear extension to correlation, borrowed from the information theory, is the mutual information (MI). The mathematical definition of MI and its estimation will be detailed in the next section.

Feature selection necessitates to select *sets of* features. This means that it is the relevance of the sets that must be evaluated, rather than the relevance of the features in the set. Indeed evaluating individually the relevance of single features would result in similar relevances; if two highly correlated, but highly relevant too, features are contained in the original set, they will then both be selected, while selecting one would have been sufficient for the prediction or classification model. Evaluating sets of features means in other words, to be able to evaluate the relevance of a multi-dimensional variable (a vector), instead of a scalar one only. Again MI is appropriate with this respect, as detailed in the next section.

Finally, many greedy procedures are proposed in the literature. While several variants exist, they can be roughly categorized in *forward* and *backward* procedures; the former means that the set is built from scratch by adding relevant features at each step, while the latter proceeds by using the whole set of initial features and removing irrelevant ones. Both solutions have their respective advantages and drawbacks. A drawback of the forward procedure is that the initial choices (when few features are concerned) influence the final choice, and may reveal suboptimal. However, the forward procedure has an important advantage: the maximum size of the vectors (feature sets) that have to be evaluated by the criterion is equal to the final set size. In the backward approach, the maximum size is the one in the initial step, i.e. the size of the initial feature set. As it will be seen in the next section, the evaluation of the criterion is also made more difficult because of the curse of dimensionality; working in smaller space dimensions is thus preferred, what justifies the choice for a forward approach.

3   Feature Selection with Mutual Information

A prediction (or classification) model aims to reduce the uncertainty on the output, the dependent variable. As mentioned in the previous section, a good criterion to evaluate the relevance of a (set of) feature(s) is nothing else than a simplified prediction model. A natural idea is then to measure the uncertainty of the output, given the fact that the inputs (independent variables) are known. The formalism below is inspired from [10].

3.1   Mutual Information Definition

A powerful formalization of the uncertainty of a random variable is Shannon's entropy. Let X and Y be two random variables; both may be multidimensional (vectors). Let $\mu_X(x)$ and $\mu_Y(y)$ the (marginal) probability density functions (pdf) of X and Y, respectively, and $\mu_{X,Y}(x,y)$ the joint pdf of the (X,Y) variable. The entropies of X and of Y, which measures the uncertainty on these variables, are defined respectively as

$$H(X) = - \int \mu_X(x) \log \mu_X(x) dx, \tag{1}$$

$$H(Y) = - \int \mu_Y(y) \log \mu_Y(y) dy. \tag{2}$$

If Y depends on X, the uncertainty on Y is reduced when X is known. This is formalized through the concept of conditional entropy:

$$H(Y|X) = - \int \mu_X(x) \int \mu_Y(y|X = x) \log \mu_Y(y|X = x) dy dx. \tag{3}$$

The Mutual Information (MI) then measures the reduction in the uncertainty on Y resulting from the knowledge of X:

$$MI(X,Y) = H(Y) - H(Y|X). \tag{4}$$

It can easily be verified that the MI is symmetric:

$$MI(X,Y) = MI(Y,X) = H(Y) - H(Y|X) = H(X) - H(X|Y); \tag{5}$$

it can be computed from the entropies:

$$MI(X,Y) = H(X) + H(Y) - H(X,Y), \tag{6}$$

and is equal to the Kullback-Leibler divergence between the joint pdf and the product of the marginal pdfs:

$$I(X,Y) = \int \int \mu_{X,Y}(x,y) \log \frac{\mu_{X,Y}(x,y)}{\mu_X(x)\mu_Y(y)}. \tag{7}$$

In theory as the pdfs $\mu_X(x)$ and $\mu_Y(y)$ may be computed from the joint one $\mu_{X,Y}(x,y)$ (by integrating over the second variable), one only needs $\mu_{X,Y}(x,y)$ in order to compute the MI between X and Y.

3.2   Mutual Information Estimation

According to the above equations, the estimation of the MI between X and Y may be carried out in a number of ways. For instance, equation (6) may be used after the entropies of X, Y and X, Y are estimated, or the Kullback-Leibler divergence between the pdfs may be used as in equation (7).

The latter solution necessitates to estimate the pdfs from the know sample (the measured data). Many methods exist to estimate pdfs, including histograms and kernel-based approximations (Parzen windows), see e.g. [11]. However, these approaches are inherently restricted to low-dimensional variables. If the dimension of X exceeds let's say three, histograms and kernel-based pdf estimation requires a prohibitive number of data; this is a direct consequence of the curse of dimensionality and the so-called empty space phenomenon. However, as mentioned in the previous section, the MI will have to be estimated on sets of features (of increasing dimension in the case of a forward procedure). Histograms and kernel-based approximators become rapidly inappropriate for this reason.

Although not all problems related to the curse of dimensionality are solved in this way, it appears that directly estimating the entropies is a better solution, at least if an efficient estimator is used. Intuitively, the uncertainty on a variable is high when the distribution is flat and small when it has high peaks. A distribution with peaks means that neighbors (or successive values in the case of a scalar variable) are very close, while in a flat distribution the distance between a point and its neighbors is larger. Of course this intuitive concept only applies if there is a finite number of samples; this is precisely the situation where it is needed to estimate the entropy rather than using its integral definition. This idea is formalized in the Kozachenko-Leonenko estimator for differential Shannon entropy [7]:

$$\hat{H}(X) = -\psi(K) + \psi(N) + \log c_D + \frac{D}{N} \sum_{n=1}^{N} \log \epsilon(n, K) \qquad (8)$$

where N is the number of samples x_n in the data set, D is the dimensionality of X, c_D is the volume of a unitary ball in a D-dimensional space, and $\epsilon(n, K)$ is twice the distance from x_n to its K-th neighbour. K is a parameter of the estimator, and ψ is the digamma function given by

$$\psi(t) = \frac{\Gamma'(t)}{\Gamma(t)} = \frac{d}{dt} \ln \Gamma(t), \qquad (9)$$

with

$$\Gamma(t) = \int_0^\infty u^{t-1} e^{-u} du. \qquad (10)$$

The same intuitive idea of K-th nearest neighbor is at the basis of an estimator of the MI between X and Y. The MI is aimed to measure the loss of uncertainty on Y when X is known. In other words, this means to answer the question whether some (approximate) knowledge on the value of X may help identifying what can be the possible values for Y. This is only feasible if there exists a certain notion of continuity or smoothness when looking to Y with respect to X. Therefore, close values in X should result in corresponding close values in Y. This is again a matter of K-nearest neighbors: for a specific data point, if its neighbors in the X and Y spaces correspond to the same data, then knowing X helps in knowing Y, which reflects a high MI.

More formally, let us define the joint variable $Z = (X, Y)$, and $z_n = (x_n, y_n), 1 \leq n \leq N$ the available data. Next, we define the norm in the Z space as the maximum norm between the X and Y components; if $z_n = (x_n, y_n)$ and $z_m = (x_m, y_m)$, then

$$\|z_n - z_m\|_\infty = \max(\|x_n - x_m\|, \|y_n - y_m\|), \tag{11}$$

where the norms in the X and Y spaces are the natural ones. Then $z_{K(n)}$ is defined as the K-nearest neighbor of z_n (measured in the Z space). $z_{K(n)}$ can be decomposed in its x and y parts as $z_{K(n)} = (x_{K(n)}, y_{K(n)})$; note however that $x_{K(n)}$ and $y_{K(n)}$ are not (necessarily) the K-nearest neighbors of x_n and y_n respectively, with $z_n = (x_n, y_n)$.

Finally, we denote

$$\epsilon_n = \|z_n - z_{K(n)}\|_\infty \tag{12}$$

the distance between z_n and its K-nearest neighbor. We can now count the number $\tau_x(n)$ of points in X whose distance from x_n is strictly less than ϵ_n, and similarly the number $\tau_y(n)$ of points in Y whose distance from y_n is strictly less than ϵ_n. It can then be shown [8] that $MI(X, Y)$ can be estimated as:

$$\widehat{MI}(X, Y) = \psi(K) + \psi(N) - \frac{1}{N} \sum_{n=1}^{N} [\psi(\tau_x(n)) + \psi(\tau_y(n))]. \tag{13}$$

As with the Kozachenko-Leonenko estimator for differential entropy, K is a parameter of the algorithm and must be set with care to obtain an acceptable MI estimation. With a small value of K, the estimator has a small bias but a high variance, while a large value of K leads to a small variance but a high bias.

In summary, while the estimator (13) may be efficiently used to measure the mutual information between X and Y (therefore the relevance of X to predict Y), it still suffers from two limitations. Firstly, there is a parameter (K) in the estimator that must be chosen with care. Secondly, it is anticipated that the accuracy of the estimator will decrease when the dimension of the X space increases, i.e. along the steps of the forward procedure. These two limitations will be addressed further in this contribution.

3.3   Greedy Selection Procedure

Suppose that M features are initially available. As already mentioned in Section 2, even if the relevance criterion was well-defined and easy to estimate, it is usually not possible to test all $2^M - 1$ non-empty subsets of features in order to select the best one. There is thus a need for a greedy procedure to reduce the search space, the aim being to have a good compromise between the computation time (or the number of tested subsets) and the potential usefulness of the considered subsets. In addition, the last limitation mentioned in the previous subsection gives the preference to greedy search avoiding subsets with a too large number of features.

With these goals in mind, it is suggested to use a simple forward procedure. The use of a backward procedure (starting from the whole set of M features) is not considered to avoid having to evaluate mutual information on M-dimensional vectors.

The forward search consists first in selecting the feature that maximizes the mutual information with the output Y:

$$X_{s_1} = \arg \max_{X_j, 1 \leq j \leq M} \widehat{MI}(X_j, Y). \tag{14}$$

Then in step t $(t \geq 2)$, the t-th features is selected as

$$X_{s_t} = \arg \max_{X_j, 1 \leq j \leq M, j \notin \{s_1, s_2, \ldots, s_{t-1}\}} \widehat{MI}(\{X_{s_1}, X_{s_2}, \ldots, X_{s_{t-1}}, X_j\}, Y). \tag{15}$$

Selecting features incrementally as defined by equations (14) and (15) makes the assumption that once a feature is selected, it should remain in the final set. Obviously, this can lead to a suboptimal solution: it is not because the first feature (for example) is selected according to equation (14) that the optimal subset necessarily contains this feature. In other words, the selection process may be stuck in a local minimum. One way to decrease the probability of being stuck in a local minimum is to consider the removing of a single feature at each step of the algorithm. Indeed, there is no reason that a selected feature (for example the first one according to equation (14)) belongs to the optimal subset. Giving the possibility to remove a feature that has become useless after some step of the procedure is thus advantageous, while the increased computational cost is low. More formally, the feature defined as

$$X_{s_d} = \arg \max_{X_j, 1 \leq j \leq t-1} \widehat{MI}(\{X_{s_1}, X_{s_2}, \ldots, X_{s_{j-1}}, X_{s_{j+1}}, \ldots, X_{s_t}\}, Y) \tag{16}$$

is removed if

$$\widehat{MI}(\{X_{s_1}, \ldots, X_{s_{d-1}}, X_{s_{d+1}}, \ldots, X_{s_t}\}, Y) > \widehat{MI}(\{X_{s_1}, \ldots, X_{s_t}\}, Y). \tag{17}$$

Of course, the idea or removing features if the removal leads to an increased MI may be extended to several features at each step. However, this is nothing else than extending the search space of subsets. The forward-backward procedure consisting in considering the removal of only one feature at each step is thus a good compromise between expected performances and computational cost.

Though the above suggestion seems to be appealing, and is used in many state-of-the-art works, it is theoretically not sound. Indeed, it can easily be shown that the mutual information can only increase if a supplementary variable is added to a set [2]. The fact that equation (17) may hold in practice is only due to the fact that equations (16) and (17) involve estimations of the MI, and not the theoretical values. The question is then whether it is legitimate to think that condition (17) will effectively lead to the removal of unnecessary features, or if this condition will be fulfilled by chance, without a sound link to the non-relevance of the removed features.

Even if the backward procedure is not used, the same problem appears. In theory indeed, if equation (15) is applied repeatedly with the true MI instead of the estimated one, the MI will increase at each step. There is thus no stopping criterion, and without additional constraint the procedure will result in the full set of M initial features! The traditional way is then to stop when the estimation of the MI begins to decrease. This leads to the same question whether the decrease of the estimated value is only due to a bias or noise in the estimator, or has a sound link to the non or low relevance of a feature.

3.4   The Problems to Solve

To conclude this section, coupling the use of an estimator of the mutual information, even if this estimator is efficient, to a greedy procedure raises several questions and problems. First, the estimator includes (as any estimator) a smoothing parameter that has to be set with care. Secondly, the dimension of the vectors from which a MI has to be estimated may have an influence on the quality of the estimation. Finally, the greedy procedure (forward, or forward-backward) needs a stopping criterion. In the following section, we propose to solve all these issues together by the adequate use of resampling methods. We also introduce an improvement to the concept of mutual information, when used to measure the relevance of a (set of) features.

4   Improving the Feature Selection by MI

In this section, we first address the problem of setting the smoothing parameter in the MI estimator, by using resampling methods. Secondly, we show how using the same resampling method provides a natural and sound stopping criterion for the greedy procedure. Finally, we show how to improve the concept of MI, by introducing a conditional redundancy concept.

4.1   Parameter Setting in the MI Estimation

The estimator defined by equation (13) faces a classical bias/variance dilemma. While the estimator is known to be consistent (see [6]), it is only asymptotically unbiased and can therefore be biased on a finite sample. Moreover, as observed in [8], the number of neighbors K acts as a smoothing parameter for the estimator: a small value of K leads to a large variance and a small bias, while a large value of K has the opposite effects (large bias and small variance).

Choosing K consists therefore in balancing the two sources of inaccuracy in the estimator. Both problems are addressed by resampling techniques. A cross-validation approach is used to evaluate the variance of the estimator while a permutation method provide some baseline value of the mutual information that can reduce the influence of the bias. Then K is chosen so as to maximize the significance of the high MI values produced by the estimator.

The first step of this solution consists in evaluating the variance of the estimator. This is done by producing "new" datasets drawn from the original dataset $\Omega = \{(x_n, y_n)\}, 1 \leq n \leq N$. As the chosen estimator strongly overestimates the mutual information when submitted replicated observations, the subsets cannot be obtained via random sampling with replacement (i.e. bootstrap samples), but are on the contrary strict subsets of Ω. We use a cross-validation strategy: Ω is split randomly into S non-overlapping subsets $U_1, \ldots, U_S$ of approximately equal sizes that form a partition of Ω. Then S subsets of Ω are produced by removing a U_s from Ω, i.e. $\Omega_s = \Omega \setminus U_s$. Finally, the MI estimator is applied on each Ω_s for the chosen variables and a range of values to explore for K. For a fixed value of K the S obtained values $\widehat{MI}_s(X, Y)$ ($1 \leq s \leq S$) provide a way to estimate the variance of the estimator.

The bias problem is addressed in a similar way by providing some reference value for the MI. Indeed if X and Y are independent variables, then $MI(X, Y) = 0$. Because of the bias (and variance) of the estimator, the estimated value $\widehat{MI}(X, Y)$ has no reason to be equal to zero (it can even be negative, whereas the mutual information is theoretically bounded below by 0). However if some variables X and Y are known to be independent, then the mean of $\widehat{MI}(X, Y)$ evaluated via a cross-validation approach provides an estimate of the bias of the estimator. In practice, given two variables X and Y known through observations $\Omega = \{(x_n, y_n)\}, 1 \leq n \leq N$, independence is obtained by randomly permuting the y_n without changing the x_n. Combined with the cross-validation strategy proposed above, this technique leads to an estimation of the bias of the estimator. Of course, there is no particular reason for the bias to be uniform: it might depend on the actual value of $MI(X, Y)$. However, a reference value if needed to obtain an estimate and the independent case is the only one for which the true value of the mutual information is known. The same X and Y as those used to calculate $\widehat{MI}_s(X, Y)$ should be of course be used, in order to remove from the bias estimation a possible dependence on the distributions (or entropies) of X and Y; just permuting the same variables helps reducing the differences between the dependent case and the reference independent one.

The cross-validation method coupled with permutation provides two (empirical) distributions respectively for $\widehat{MI}_K(X, Y)$ and $\widehat{MI}_K(X, \pi(Y))$, where π denotes the permutation operation and where the K subscript is used to emphasises the dependency on K. A good choice of K then corresponds to a situation where the (empirical) variances of $\widehat{MI}_K(X, Y)$ and $\widehat{MI}_K(X, \pi(Y))$ and the (empirical) mean of $\widehat{MI}_K(X, \pi(Y))$ are small. Another way to formulate a similar constraint is to ask for $\widehat{MI}_K(X, Y)$ to be significantly different from $\widehat{MI}_K(X, \pi(Y))$ when X and Y are known to exhibit some dependency. The differences between the two distributions can be measured for instance via a measure inspired from Student's t-test. Let us denote μ_K (resp. $\mu_{K,\pi}$) the empirical mean of $\widehat{MI}_K(X, Y)$ (resp. $\widehat{MI}_K(X, \pi(Y))$) and σ_K (resp. $\sigma_{K,\pi}$) its empirical standard deviation.

Then the quantity

$$t_K = \frac{\mu_K - \mu_{K,\pi}}{\sqrt{\sigma_K^2 + \sigma_{K,\pi}^2}} \tag{18}$$

measures the significance of the differences between the two (empirical) distributions (if the distributions were Gaussian, a Student's t-test of difference in the means of the distributions could be conducted). Then one chooses the value of K that maximizes the differences between the distributions, i.e. the one that maximizes t_K.

The pseudo-code for the choice of K in the MI estimator is given in Table 1. In practice, the algorithm will be applied to each real valued variable that constitute the X vector and the optimization of K will be done along all the obtained t_K values. As t_K will be larger for relevant variables than for non-relevant ones, this allows to discard automatically the influence of non-relevant variables in the choice of K.

Table 1. Pseudo-code for the choice of K in the mutual information estimator

```
Inputs  Ω = {(xₙ, yₙ)}, 1 ≤ n ≤ N the dataset
        Kmin and Kmax the range where to look for the optimal K
        S the cross-validation parameter
Output  the optimal value of K

Code    Draw a random partition of Ω into S subsets U₁, ..., Uₛ
            with roughly equal sizes
        Draw a random permutation π of {1, ..., N}
        For K ∈ {Kmin, ..., Kmax}
            For s ∈ {1, ..., S}
                compute mi[s] the estimation of the mutual information
                    MI(X, Y) based on Ωₛ = Ω \ Uₛ
                compute miπ[s] the estimation of the mutual information
                    MI(X, π(Y)) based on Ωₛ = Ω \ Uₛ
                    with the permutation π applied to the yᵢ
            EndFor
            Compute μK the mean of mi[s] and σK its standard deviation
            Compute μK,π the mean of miπ[s] and σK,π its standard deviation
            Compute tK = (μK − μK,π)/√(σK²+σK,π²)
        EndFor
        Return the smallest K that minimises tK on {Kmin, ..., Kmax}
```

To test the proposed methodology, a dataset is generated as follows. Ten features $X_i, 1 \leq i \leq 10$ are generated from a uniform distribution in $[0, 1]$. Then, Y is built according to

$$Y = 10\sin(X_1 X_2) + 20(X_3 - 0.5)^2 + 10X_4 + 5X_5 + \epsilon, \tag{19}$$

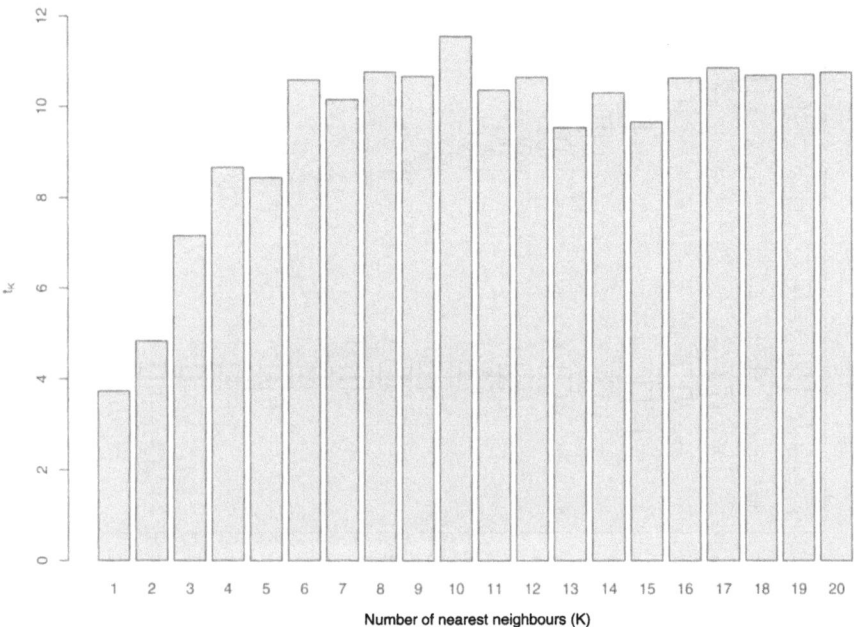

Fig. 1. Values of t_K for $MI(X_4, Y)$ (see text for details)

where ϵ is a Gaussian noise with zero mean and unit variance. Note that variables X_6 to X_{10} do not enter into equation (19); they are independent from the output Y. A sample size of 100 observations is used and the CV parameter is $S = 20$. When evaluating the MI between Y and a relevant feature (for example X_4), a t_K value is obtained for each value of K, as shown on Figure 1. Those values summarize the differences between the empirical distributions of $\widehat{MI}_K(X_4, Y)$ and of $\widehat{MI}_K(X_4, \pi(Y))$ (an illustration of the behaviour of those distributions is given in Figure 2). The largest t_K value corresponds to the smoothing parameter K that best separates the distributions in the relevant and permuted cases (in this example the optimal K is 10).

4.2   Stopping Criterion

As mentioned above, stopping the greedy forward or forward-backward procedure when the estimated MI decreases is not sound or theoretically justified. A better idea is to measure whether the addition of a feature to the already selected set increases significantly the MI, compared to a situation where a non-relevant feature is added, again in the same settings i.e. keeping the same distribution for the potentially relevant variable and the non-relevant one.

This problem is similar to the previous one. Given a subset of already selected variables S, a new variable X_{st} is considered significant if the value of $MI(S \cup X_{st}, Y)$ significantly differs from the values generated by $MI(S \cup \pi(X_{st}), Y)$,

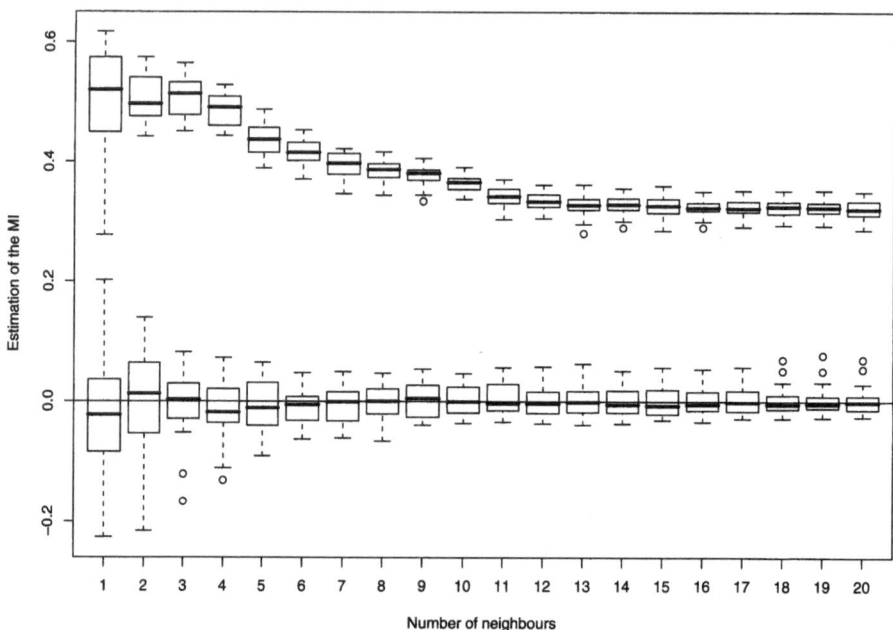

Fig. 2. Boxplots for the distributions of $\widehat{MI}_K(X_4, Y)$ (top) and of $\widehat{MI}_K(X_4, \pi(Y))$ (bottom) as a function of K

where π is a random permutation. In practice, one generates several random permutation and counts the number of times that $MI(S \cup \pi(X_{st}), Y)$ is higher than $MI(S \cup X_{st}, Y)$. This gives an estimate of the p-value of $MI(S \cup X_{st}, Y)$ under the null hypothesis that X_{st} is independent from (S, Y). A small value means that the null hypothesis should be rejected and therefore that X_{st} brings significant new information about Y. The pseudo-code for the proposed algorithm is given in Table 2.

It should be noted that the single estimation of $MI(S \cup X_{st}, Y)$ could be replaced by a cross-validation based estimate of the distribution of this value. The same technique should then be used in the estimation of the distribution of $MI(S \cup \pi(X_{st}), Y)$.

To illustrate this method, 100 datasets are randomly generated according to equation (19). For each dataset, the optimal value of K for the MI estimator is selected according to the method proposed in Section 4.1. Then a forward procedure is applied and stopped according to the method summarized in Table 2 (with a significance threshold of 0.05 for the p-value). As it can be seen from Table 3, in most cases 4 or 5 relevant features are selected by the procedure (5 is the expected number, as X_6 to X_{10} are not linked with Y). Without resampling, by stopping the forward procedure at the maximum of mutual information, in most cases only 2 (45 cases) and 3 (33 cases) features are selected. This is a consequence of the fact that when looking only to the value of the estimated

Table 2. Pseudo-code for the choice of the stopping criterion for the greedy procedure

```
Inputs  Ω = {(xₙ, yₙ)}, 1 ≤ n ≤ N the dataset
        P the number of permutations to compute
        the subset S of currently selected variables
        the candidate variable X_st
Output  a p-value for the hypothesis that the variable is useless
Code    Compute ref the value of MI(S ∪ X_st, Y)
        Initialise out to 0
        For p ∈ {1, ..., P}
            Draw a random permutation π_p of {1, ..., N}
            If MI(S ∪ π_p(X_st), Y) ≥ ref then
                increase out by 1
            EndIf
        EndFor
        Return out/P
```

Table 3. Number of selected features

Number of features	1	2	3	4	5	6
Percentage	0	1	12	52	29	6

MI at each step, the estimation is made in spaces of increasing dimension (the dimension of X is incremented at each step). It appears that in average the estimated MI decreases with the dimension, making irrelevant the comparison of MI estimations with feature vectors of different dimensions.

More experiments on the use of resampling to select K and to stop the forward procedure may be found in [5].

4.3   Clustering by Rank Correlation

In some problems and applications, the number of features that are relevant for the prediction of variable Y may be too large to afford the above described procedures. Indeed, as detailed in the previous sections, the estimator of mutual information will fail when used on too high-dimensional variables, despite all precautions that are taken (using an efficient estimator, avoiding a backward procedure, using estimator results on a comparative basis rather than using the rough values, etc.).

In this case, another promising direction is to cluster features instead of selecting them [9,3]. Feature clustering consists in grouping features in natural clusters, according to a similarity measure: features that are similar should be grouped in a single cluster, in order to elect a single representative from the latter. This is nothing else than applying to features the traditional notion of clustering usually applied to objects. For example, all hierarchical clustering methods can be used, the only specific requirement being to define a measure of similarity between features. Once the measure of similarity is defined, the clustering consists in selecting the two most similar features and replacing them by a representative. Next, the procedure is repeated on the remaining initial features and representatives.

The advantage of feature clustering with respect to the procedure based on the mutual information between a group of features and the output, as described above, is that the similarity measure is only used on two features (or representatives) at each iteration. Therefore the problems related to the increasing dimensionality of the feature sets is completely avoided. The reason behind this advantage is that in the first case the similarity is measured between a set of features and the output, while in the clustering the similarity is measured between features (or their representatives) only. Obviously, the drawback is that the variable Y to predict is no more taken into account.

In order to remedy to this last problem, a new conditional measure of similarity between features is introduced. Simple correlation or mutual information between features could be used, but will not take the information from Y into account. However, based on the idea of Kraskov's estimator of Mutual Information, one can define a similarity measure that takes Y into account, as follows [4].

Let X_1 and X_2 the two features whose similarity should be measured. The idea is to measure if X_1 and X_2 contribute similarly, or not, to the prediction of Y. Let $\Omega = z_n = (x_{1n}, x_{2n}, y_n), 1 \leq n \leq N$ the sample set; other features than X_1 and X_2 are discarded from the notation for simplicity. For each element n, we search for the nearest neighbor according to the Euclidean distance in the joint (X_1, Y) space. Let we denote this element by its index m. Then, we count the number $c1_n$ of elements that are closer from element n than element m, taking only into account the distance in the X_1 space. Figure 3 shows such elements. $c1_n$ is a measure of the number of local false neighbors, i.e. elements that are neighbors according to X_1 but not according to (X_1, Y). If this number if high, it means that element n can be considered as a local outlier in the relation between X_1 and Y.

The process is repeated for all elements n in the sample set, and resulting $c1_n$ values are concatenated in a N-dimensional vector $C1$. Next, the same procedure is applied with feature X_2 instead of X_1; resulting $c2_n$ values are concatenated in vector $C2$.

If features X_1 and X_2 carry the same information to predict Y, vectors $C1$ and $C2$ will be similar. On the contrary, if they carry different yet complementary information, vectors $C1$ and $C2$ will be quite different. Complementary information can be for example that X_1 is useful to predict Y in a part of its range, while X_2 plays a similar role in another part of the range. As the $c1_n$ and $c2_n$ vectors contain local information in the (X_1, Y) (respectively (X_2, Y)) relation, vectors $C1$ and $C2$ will be quite different in this case. For these reasons, the correlation between $C1$ and $C2$ is a good indicator of the similarity between X_1 and X_2 when these features are used to predict Y. This is the similarity measure that is used in the hierarchical feature clustering algorithm.

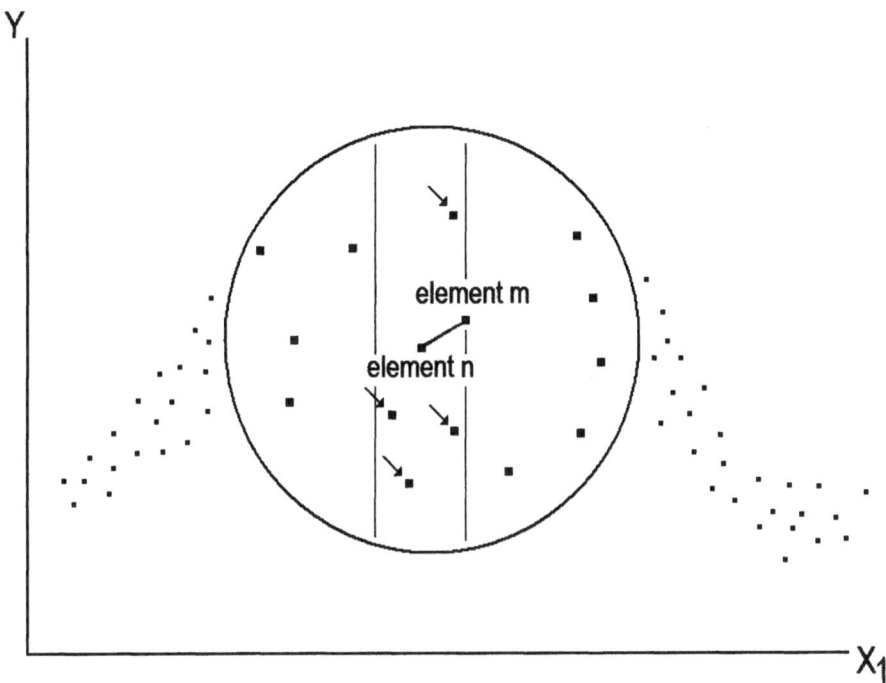

Fig. 3. Identification of neighbors in X_1 that are not neighbors in (X_1, Y)

In order to illustrate this approach, it is applied to two feature clustering problems where the number of initial features is large. Analysis of (infrared) spectra is a typical example of such problem. The first dataset, Wine citewine, consists in 124 near-infrared spectra of wine samples, for which the concentration in alcohol has to be predicted from the spectra. Three outliers are removed, and 30 spectra are kept aside for test. The second dataset is the standard Tecator benchmark [1]; it consists of 215 near-infrared spectra of meat samples, 150 of them being used for learning and 65 for test. The prediction model used for the experiments is Partial Least Squares Regression (PLSR); the number of components in the PLSR model is set by 4-fold cross-validation on the training set. Three experiments are conducted on each set: a PLSR model on all features, a PLSR model on traditional clusters built without using Y, and a PLSR model built on clusters defined as above. The results are shown in Table 4; the Normalized Mean Square Error (NMSE) on the test set is given, together with the number of features or clusters.

In both cases, the clustering using the proposed method (last column) performs better than a classical feature clustering, or no clustering at all. In the Tecator experiment, the advantage in terms of performances with respect to the non-supervised clustering is not significant; however, in this case, the number of resulting clusters is much smaller in the supervised case, which reaches the

Table 4. Results of the feature clustering on two spectra datasets

	without clustering	clustering without Y	clustering with Y
Wine	$NMSE = 0.00578$	$NMSE = 0.0111$	$NMSE = 0.00546$
	256 features	19 clusters	33 clusters
Tecator	$NMSE = 0.02658$	$NMSE = 0.02574$	$NMSE = 0.02550$
	100 features	17 clusters	8 clusters

fundamental goal of feature selection, i.e. the ability to build simple, inter-pretable models.

5   Conclusion

Feature selection in supervised regression problems is a fundamental preprocess-ing step. Feature selection has two goals. First, similarly to other dimension reduction techniques, it is aimed to reduce the dimensionality of the problem without significant loss of information, therefore acting against the curse of di-mensionality. Secondly, contrarily to other approaches where new variables are built from the original features, feature selection helps to interpret the resulting prediction model, by providing a relevance measure associated to each original feature.

Mutual Information (MI), a concept borrowed from information theory, can be used for feature selection. The MI criterion is used to test the relevance of subsets of features with respect to the prediction task, in a greedy proce-dure. However, in practise, the MI theoretical concept needs to be estimated. Even if efficient estimators exist, they still suffer from two drawbacks: their per-formances decrease with the dimension of the feature subsets, and they need to set a smoothing parameter (for example K in a K-nearest neighbors based estimator).

In addition, when embedded in a forward selection procedure, the MI does not provide any stopping criterion, at least in theory. Standard practice to stop the selection when the estimation of the MI begins to decrease exploits in fact a limitation of the estimator itself, without any guarantee that the algorithm will indeed stop when no further feature has to be added.

This chapter shows how to cope with these three limitations. It shows how using resampling and permutations provides first a way to compare MI values on a sound basis, and secondly a stopping criterion in the forward selection process.

In addition, when the number of relevant features is high, there is a need to avoid using MI between feature sets and the output, because of the too high dimension of the feature sets. It is also shown how to cluster features by a similarity criterion used on single features. The proposed criterion measures whether two features contribute identically or in a complementary way to the prediction of Y; the measure is thus supervised by the prediction task.

These methodological proposals are shown to improve the results of a feature selection process using similarity measures based on Mutual Information.

References

1. Borggaard, C., Thodberg, H.: Optimal minimal neural interpretation of spectra. Analytical Chemistry 64, 545–551 (1992)
2. Cover, T., Thomas, J.: Elements of Information Theory. Wiley, New York (1991)
3. Dijck, G.V., Hulle, M.M.V.: Speeding up the wrapper feature subset selection in regression by mutual information relevance and redundancy analysis. In: ICANN 2006: International Conference in Aritificial Neural Networks, Athens, Greece, September 2006, pp. 31–40 (2006) (submitted)
4. François, D., Krier, C., Rossi, F., Verleysen, M.: Estimation de redondance pour le clustering de variables spectrales. In: Agrostat 2008, 10èmes journées Européennes Agro-industrie et Méthodes statistiques, pp. 55–61. Louvain-la-Neuve, Belgium (2008)
5. François, D., Rossi, F., Wertz, V., Verleysen, M.: Resampling methods for parameter-free and robust feature selection with mutual information. Neurocomputing 70(7-9), 1265–1275 (2007)
6. Goria, M.N., Leonenko, N.N., Mergel, V.V., Inverardi, P.L.N.: A new class of random vector entropy estimators and its application in testing statistical hypotheses. Journal of Nonparametric Statistics 17(3), 277–297 (2005)
7. Kozachenko, L.F., Leonenko, N.N.: Sample estimate of entropy of a random vector. Probl. Inf. Transm. 23, 95–101 (1987)
8. Kraskov, A., Stögbauer, H., Grassberger, P.: Estimating mutual information. Physical Review E 69, 066138 (2004)
9. Krier, C., François, D., Rossi, F., Verleysen, M.: Feature clustering and mutual information for the selection of variables in spectral data. In: ESANN 2007, European Symposium on Artificial Neural Networks Advances in Computational Intelligence and Learning, pp. 157–162. Bruges, Belgium (2007)
10. Rossi, F., Lendasse, A., François, D., Wertz, V., Verleysen, M.: Mutual information for the selection of relevant variables in spectrometric nonlinear modelling. Chemometrics and Intelligent Laboratory Systems 2(80), 215–226 (2006)
11. Scott, D.: Multivariable Density Estimation: Theory, Practice, and Visualization. Wiley, New-York (1992)

Unleashing Pearson Correlation for Faithful Analysis of Biomedical Data

Marc Strickert[1], Frank-Michael Schleif[2], Thomas Villmann[2], and Udo Seiffert[3]

[1] Data Inspection Group, Leibniz Institute of Plant Genetics and
Crop Plant Research (IPK) Gatersleben, Germany
`stricker@ipk-gatersleben.de`
[2] Research group Computational Intelligence, University of Leipzig, Germany
`{schleif@informatik,thomas.villmann@medizin}.uni-leipzig.de`
[3] Biosystems Engineering, Fraunhofer Institute for Factory Operation and
Automation (IFF), Magdeburg, Germany
`udo.seiffert@iff.fraunhofer.de`

Abstract. Pearson correlation is one of the standards for comparisons in biomedical analyses, possessing yet unused potential. Substantial value is added by transferring Pearson correlation into the framework of adaptive similarity measures and by exploiting properties of the mathematical derivatives. This opens access to optimization-based data models applicable in tasks of attribute characterization, clustering, classification, and visualization. Modern high-throughput measuring equipment creates high demand for analysis of extensive biomedical data including spectra and high-resolution gel-electrophoretic images. In this study cDNA arrays are considered as data sources of interest. Recent computational methods are presented for the characterization and analysis of these huge-dimensional data sets.

Keywords: high-dimensional data mining, feature rating, clustering, data visualization, parametric correlation measure.

1   Introduction

Massive data sets with a high number of samples and/or attributes create interesting challenges in *de novo* data analysis. Particularly, transcriptome and proteome studies with their high-throughput biomedical screening devices, such as mass spectrometers, gene expression arrays, or protein gels, generate thousands of data points in parallel for which accurate processing is needed [2]. Typical analysis tasks comprise clustering and classification as well as the characterization of data attributes for dimension reduction, biomarker construction, and visualization.

For its intuitive relation to the ordinary physical space, Euclidean distance is probably the most widely spread approach to the comparison of high-dimensional data vectors. Its predominant presence in textbooks certainly results from the good interpretability of the Euclidean space. Many researchers have recognized

M. Biehl et al.: (Eds.): Similarity-Based Clustering, LNAI 5400, pp. 70–91, 2009.

the need for alternative ways of data comparison and they employ more general Minkowski metrics, correlation similarity or other elaborate measures, for example, targeting comparison of text data [19,38,40]. In biomedical analysis, correlation-based analysis plays an important role, because usually linear but also more general dependencies between observations can be characterized by correlation coefficients. In addition to regression analysis, correlation with its built-in mean centering and variance normalization is beneficial in pattern matching tasks. Accounting for profiles of sequential data values, in some cases correlation can be used as a poor man's approach to functional processing, for example, in co-expression analysis of temporal gene expression data [27] and in comparisons of chromatographic data series [26].

The need for fully correlation-driven methods is increasingly noticed in the data analysis community. For example, a correlation-based version of linear discriminant analysis (LDA), called correlation discriminant analysis (CDA), has been recently proposed [21]. Vector quantization approaches [11] such as learning vector quantization or self-organizing maps [16] were originally designed for the Euclidean distance. These methods might not yield optimum performance if only the distance metric is replaced by a similarity measure like correlation. Although this is usually done, further refinement of their update mechanisms in accordance with the chosen measure is also advisable.

The Euclidean paradigm is yet strong enough to effectively prevent asking, whether concepts like the widely used first two statistical moments, the mean and variance, are really applicable to the space induced by an alternatively chosen measure. For example, popular methods like the k-means clustering algorithm [12] will not work optimally when the arithmetic center of gravity does not coincide with the point of maximum similarity to the surrounding data which is determined by the similarity measure of custom choice. An illustration regarding this first statistical moment, the mean, will be given in comparison to Pearson correlation later. Another issue of discussion will concern the second statistical moment: since classical variance is not always compatible with the underlying data measure, common methods like principal component analysis (PCA) will produce misleading results about directions of maximum variance in structurally non-Euclidean data spaces. A counterpart to statistical variance will be also presented for Pearson correlation.

Here, after revisiting the most common concepts of correlation (Section 2) and presenting derivatives of an adaptive version of Pearson correlation (Section 2.3), we focus on correlation-based attribute rating (Section 3), space partitioning using vector quantization (Section 4), and a visualization method (Section 5). In addition to the variance-related approach to attribute rating (Section 3.1), supervised attribute relevance determination using cross comparison (SARDUX) is presented with an adaptive correlation measure for labeled data (Section 3.2). Faithful space partitioning is obtained by neural gas vector quantization [23] using correlation (NG-C) [33]. Data visualization is realized by correlation-based multi-dimensional scaling for high-throughput data (HiT-MDS) [32].

Table 1. List of commonly used symbols

$\mathbf{X}$	data set
$\mathbf{D}$; $\hat{\mathbf{D}}$	data similarity matrix; reconstructed similarity matrix
$\mathbf{x}$; $\mathbf{x}^j$; x_j^k	data vector; j-th data vector; k-th component of j-th data vector
$\boldsymbol{w}$	comparison vector or data prototype vector
$\mathbf{x}^z$	z-score transformed data vector
$\langle \mathbf{x}, \boldsymbol{w} \rangle$	scalar product between $\mathbf{x}$ and $\boldsymbol{w}$
λ; λ_i	parameter vector; i-th component of adaptive Pearson correlation
$\mathsf{r}_\lambda (\mathbf{x}, \boldsymbol{w})$	adaptive Pearson correlation between $\mathbf{x}$ and $\boldsymbol{w}$
d	data dimension
n	number of instances in data set
γ	learning rate used in gradient descent
μ; σ	average value; variance, neighborhood range
s	stress function

2   Correlation Measures

In general, correlation characterizes the strictness of dependence of attributes in two vectors of identical dimensionality. 'The more quantity in one vector attribute, the more of it in the same attribute in the other vector' is an example of positive correlation, while 'the more, the less' and 'the less, the more' indicate negative correlation. Large absolute correlation coefficients, though, are no guarantee that two observations really influence one another, because they might be caused by spurious dependence, either mediated by a hidden factor controlling both, or simply by chance. Therefore the frequently found term 'measure of association' might be a misleading synonym of correlation.

Different types of correlation can be specified. For brevity, we refer to existing literature [20] and outline only the three most frequent ones. These are $\langle 1 \rangle$ Kendall's Tau [τ], for analyzing the occurrences of positive and negative signs in comparisons of potentially ordinal components in two data vectors; $\langle 2 \rangle$ Spearman rank correlation, comparing rank orderings, i.e. simultaneously monotonic relationships, of the entries of two vectors; and $\langle 3 \rangle$ Pearson correlation which is a measure of linear correlation between real-value entries of two vectors. All three measures yield numerical values in a range between -1, (negative correlation) and +1 (positive correlation); values around zero indicate uncorrelatedness. Recent studies propose measures for combining the desired quantification properties of these three types of correlation [5,39]. For its widespread use and its beneficial mathematical structure we will focus on Pearson correlation in later sections.

2.1   Kendall's Tau [τ]

The Kendall coefficient τ measures the strength of the common tendency of two d-dimensional vectors $\mathbf{x} = (x_k)_{k=1...d}$ and $\boldsymbol{w} = (w_k)_{k=1...d}$ in a very direct manner. Data pairs (x_i, w_i) and (x_j, w_j) are considered; if both $x_i \geq x_j$ and $w_i \geq w_j$, or both $x_i \leq x_j$ and $w_i \leq w_j$, the pair is counted as concordant, else as discordant. The number of concordant pairs is C, the number of discordant pairs is D, for $i < j$ in both cases. Then

$$\tau = \frac{C - D}{d \cdot (d - 1)/2}$$

describes the amount of bias towards concordant or discordant occurrences, finally normalized by the effective number of component pairs. Applications of Kendall's τ range from quantification of autocorrelation structure in sequential data [10] to associative studies in protein data [6].

For its fundamental counting statistics, for its applicability to ordinal data, and for its easy interpretation, Kendall's τ might be considered as favorable characterization of correlation. However, the computing complexity is $\mathcal{O}(d^2)$, i.e. quadratic in the number d of data dimensions. Also, values of τ change in discrete steps as data relationships change; this makes it difficult to use in optimization scenarios, such as for the goal of adaptive data representation discussed in this work.

2.2   Spearman Rank Correlation [ρ]

Another non-parametric correlation measure is obtained by calculating the normalized squared Euclidean distance of the ranks of x_k and w_k according to

$$\rho(\mathbf{x}, \mathbf{w}) = 1 - \frac{6}{d \cdot (d^2 - 1)} \cdot \sum_{k=1}^{d} \left(\mathrm{rnk}(x_k) - \mathrm{rnk}(w_k) \right)^2 .$$

Ranks of real values c_k are defined by $\mathrm{rnk}(c_k) = |\{c_i < c_k , i = 1 \ldots d\}|$, which can be easily derived from the ordering index (minus one) after an ascending sorting operation. This induces a common computing complexity of $\mathcal{O}(d \cdot \log(d))$. Equivalently, data can be turned into ranks in a preprocessing step and then be subject to the Pearson correlation coefficient according to the next paragraph. Spearman rank correlation has many applications, among which the analysis of gene expression data is mentioned [3].

Spearman correlation has got the interesting property that a conversion of a non-linear ordinal data space into a rank-based Euclidean space takes place. Replacing vector entries by their ranks leads to a compression of outliers and to a magnification of close values, which, in the absence of ties, results in a uniform distribution with unit spacing and invariant statistical moments of the data vectors. In case of a low noise ratio, this simple conversion is a robust preprocessing step for getting standardized discriminations in Euclidean space. Thus, correlation analysis can be realized by Euclidean methods after data transformation. If attribute-specific metric adaptation is desired, though, Spearman correlation cannot be used in a continuous optimization framework, because the potentially required re-ranking is a mathematically non-differentiable operation.

2.3   Extended Pearson Correlation [r_λ]

In the majority of publications, linear data dependencies are characterized by values of Pearson correlation. This measure is connected to regression analysis of scattered real-valued data in terms of the residual sum of squared distances to the fitted straight line. Complementary to its central role in standard statistics, (cross-)correlation between data streams is an important concept in signal processing, as for example, for correlation optimized warping in alignment of chromatogram spectra [26]. For versatile usability, the Pearson correlation coefficient between vectors $\mathbf{x}$ and $\boldsymbol{w}$ is defined with additional parameter vector $\boldsymbol{\lambda} = (\lambda_i)_{i=1...d}$ as:

$$r_\lambda(\mathbf{x}, \boldsymbol{w}) = \frac{\sum_{i=1}^{d} \lambda_i \cdot (x_i - \mu_{\mathbf{x}}) \cdot \lambda_i \cdot (w_i - \mu_{\boldsymbol{w}})}{\sqrt{\left(\sum_{i=1}^{d} \lambda_i^2 \cdot (x_i - \mu_{\mathbf{x}})^2\right) \cdot \left(\sum_{i=1}^{d} \lambda_i^2 \cdot (w_i - \mu_{\boldsymbol{w}})^2\right)}} =: \frac{\mathscr{B}}{\sqrt{\mathscr{C} \cdot \mathscr{D}}} . \quad (1)$$

The traditional textbook notation of Pearson correlation $r = r_1$ is obtained for a choice of $\boldsymbol{\lambda} = \mathbf{1} \equiv \lambda_i = 1, i = 1 \ldots d$, but the adaptive parameter vector $\boldsymbol{\lambda}$ introduced for expressing individual attribute scalings will become useful in later considerations concerning an adaptive data measure. The shortcut notation of the fraction with canonic link of $\mathscr{B}, \mathscr{C}$ and $\mathscr{D}$ to its numerator and denominator constituents will simplify upcoming derivations.

Pearson correlation provides invariance to rescalings of whole data vectors by a common factor and to common additive component offsets, such as from homogeneous background signals. This feature makes Pearson correlation very suitable for the analysis of data from individually calibrated biomedical measuring devices [25]. The invariance feature of Pearson correlation results from implicit data normalization in Eqn. 1: the denominator normalizes the covariance $\mathscr{B}$ of mean-centered $\mathbf{x}$ and $\boldsymbol{w}$ by their individual standard deviations, expressed by square roots of $\mathscr{C}$ and $\mathscr{D}$.

Extreme values of $r_\lambda(\mathbf{x}, \boldsymbol{w}) = \pm 1$ indicate that component pairs (x_i, w_i) are located on lines with positive or negative slopes, respectively; absence of correlation is indicated by $r_\lambda(\mathbf{x}, \boldsymbol{w}) = 0$. One must be aware, though, that Pearson correlation is an ambiguous measure of linear dependence between data vectors. This is illustrated in Fig. 1 where four obviously different data sets, known as Anscombe's quartet, with the same correlation value of $r = 0.82$ are shown [1].

For those extreme cases it is pragmatic to state that the use of Pearson correlation in medicine and biology is, in the first place, justified by the its wide spreading. Yet, natural data sources will lead to more common situations like the one depicted in the very left panel of Fig. 1.

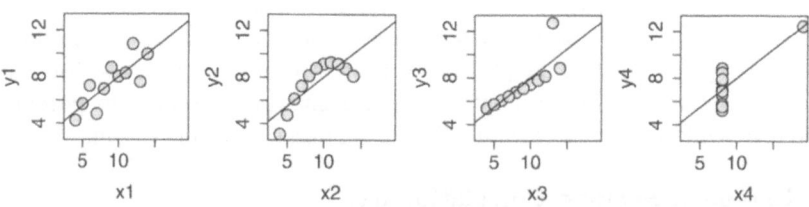

Fig. 1. Anscombe's quartet. Four data sets with same Pearson correlation coefficients of $r = 0.82$ between paired variables x and y. All regression lines are $y = 3 + 0.5 \cdot x$.

Optimization procedures will probably not be affected by artificial configurations like those mentioned in Fig. 1, which might occur as initial or transient states only. Instead, some ambiguity reflected by extra degrees of freedom during optimization helps finding optimal solutions: while in Euclidean sense, the optimal match (identity) requires the scatter plot of two data vectors to fall onto the line with unit slope, maximum correlation similarity is reached whenever the scatter constitutes a straight line with arbitrary positive slope.

Despite higher stringency for Euclidean matches between two data vectors, Pearson correlation is directly related to the Euclidean distance of z-score transformed data $\mathbf{x}^z = (\mathbf{x} - \mu_\mathbf{x})/\sqrt{\text{var}(\mathbf{x})}$. This vector operation discards the mean value of $\mathbf{x}$ and yields unit variance. For z-score transformed vectors $\mathbf{x}^z$ and $\mathbf{w}^z$ the correlation of $\mathbf{x}$ and $\mathbf{w}$ is expressed by the scaled scalar product $\langle \cdot, \cdot \rangle$:

$$\mathsf{r}(\mathbf{x}, \mathbf{w}) = \mathsf{r}(\mathbf{x}^z, \mathbf{w}^z) = \langle \mathbf{x}^z, \mathbf{w}^z \rangle / (d-1).$$

This notation helps linking Pearson correlation to squared Euclidean distance:

$$\mathsf{d}^2(\mathbf{x}^z, \mathbf{w}^z) = \sum_{i=1}^{d}(x_i^z - w_i^z)^2 = \langle \mathbf{x}^z, \mathbf{x}^z \rangle - 2 \cdot \langle \mathbf{x}^z, \mathbf{w}^z \rangle + \langle \mathbf{w}^z, \mathbf{w}^z \rangle$$

$$= 2 \cdot (d-1) \cdot \left(1 - \mathsf{r}(\mathbf{x}, \mathbf{w})\right)$$

$$\Leftrightarrow \quad \mathsf{r}(\mathbf{x}, \mathbf{w}) = 1 - \frac{1}{2} \cdot \mathsf{d}^2(\mathbf{x}^z, \mathbf{w}^z)/(d-1).$$

Although equivalence between r and d^2 can be achieved, it is much simpler to think of 'correlation' rather than of the 'shifted negative rescaled squared Euclidean distance of z-score transformed data'. The 'correlation' view is especially helpful, for example, in the case of dynamic data models where derivatives are calculated. As derivatives can be used to solve optimization tasks, two central types of derivatives related to data characterization and modeling are discussed in the following section.

Derivatives of Pearson correlation. In general, derivatives are used to characterize the amount of contribution of parameters to a differentiable functional term. The main idea is to analyze the response of functions to infinitely small parameter variations. This is not only essential for studying the sensitivity of constituents of complex systems, but also for finding parameter directions of maximum effect, as used in steepest gradient optimization methods. Many metrics or similarity measures are expressions that are differentiable almost everywhere and, consequently, sums of which or other differentiable wrappers inherit this beneficial property. Pearson correlation is well-defined as long as the denominator in Eqn. 1 is not zero. Singularity only occurs in degenerated cases of always identical components $x_i = c_1$, $w_i = c_2$, or $\lambda_i = 0$ for $i = 1 \ldots d$. A more detailed discussion of differentiability can be found in [30].

The extended formulation of Pearson correlation in Eqn. 1 allows to take two interesting approaches: one for adapting vectors, aiming at optimum data representations, the other for realizing an adaptive correlation measure using λ. Here, these two concepts are introduced as building-blocks based on the comparison of two vectors. Later, the presented expressions will be inserted into more complex scenarios with larger data sets. For simpler notation mean-subtracted vectors $x_k^\mu = (x_k - \mu_\mathbf{x})$ and $w_k^\mu = (w_k - \mu_\mathbf{w})$ are applied.

Optimum data representation. The derivative of Pearson correlation in Eqn. 1 w.r.t. the weight vector component w_k is given by the formula

$$\frac{\partial r_\lambda(\mathbf{x}, \mathbf{w})}{\partial w_k} = \lambda_k^2 \cdot r_\lambda(\mathbf{x}, \mathbf{w}) \cdot \left(\frac{x_k^\mu}{\mathscr{B}} - \frac{w_k^\mu}{\mathscr{D}} \right). \qquad (2)$$

In addition to non-constant vector components, required for the non-singular correlation values mentioned before, an additional prerequisite of non-orthogonal vectors with $(\mathscr{B} = \langle \mathbf{x}^\mu, \mathbf{w}^\mu \rangle) \neq 0$ is induced for $x_k^\mu \neq 0$. It is interesting to note that unlike Euclidean distance a non-symmetric contribution $\partial r_\lambda / \partial w_k \neq -\partial r_\lambda / \partial x_k$ of components to the derivative takes place even in case of $\lambda = 1$ [28].

Adaptive correlation measure. In Eqn. 1 the attribute scaling factors λ_i define a transformation that is applied to the vectors compared by the correlation measure. This parametric measure can be optimized in order to enhance or decrease separability of vectors taken from a common vector space. In the present case parameter adaptation of the similarity measure can be expressed by an equivalent rescaling of the data space, which is not possible for arbitrary parametric measures. The partial derivatives of the extended Pearson correlation measure are

$$\frac{\partial r_\lambda(\mathbf{x}, \mathbf{w})}{\partial \lambda_k} = -\lambda_k \cdot r_\lambda(\mathbf{x}, \mathbf{w}) \cdot \left(\frac{\mathscr{C}}{\mathscr{D}} \cdot (w_k^\mu)^2 + \frac{\mathscr{D}}{\mathscr{C}} \cdot (x_k^\mu)^2 - \frac{2}{B} \cdot w_k^\mu \cdot x_k^\mu \right). \qquad (3)$$

3   Attribute Rating and Adaptive Correlation

For high-dimensional data sets the characterization of attribute relevances is of high interest for several reasons. In dimension reduction tasks, the most informative attributes are kept, while the others can be ignored. If additional class labels are given, within-class variation can be canceled by optimum attribute rescaling, while between-class differences get emphasized. Applications of attribute rating range from active noise cancellation to biomarker identification.

3.1   Unsupervised Attribute Rating – Variance Analogon

Statistical variance expressed by the second central moment is tightly bound to Euclidean data spaces: for two data vectors $\mathbf{x}, \mathbf{w} \in \mathbb{R}^d$ the squared Euclidean distance d^2 and component derivatives are calculated by

$$d^2(\mathbf{x}, \mathbf{w}) = \sum_{i=1}^d (x_i - w_i)^2 \quad, \qquad \frac{\partial d^2(\mathbf{x}, \mathbf{w})}{\partial w_k} = -2 \cdot (x_k - w_k).$$

This allows to define the overall contribution of the k-th attribute for all vector pairs $\mathbf{x}^i$ and $\mathbf{x}^j$ in a data set $\mathbf{X} \in \mathbb{R}^d$ by the special quantity

$$V_k(\mathbf{X}) = \frac{1}{8 \cdot n \cdot (n-1)} \sum_{i=1}^{n} \sum_{j=1}^{n} \left(\frac{\partial \mathrm{d}(\mathbf{x}^i, \mathbf{x}^j)}{\partial x_k^j} \right)^2 . \tag{4}$$

For the squared Euclidean distance d^2 the term V_k can be transformed to

$$V_k(\mathbf{X}) = \frac{1}{2 \cdot n \cdot (n-1)} \sum_{i=1}^{n} \sum_{j=1}^{n} (x_k^i - x_k^j)^2 = \frac{1}{n-1} \sum_{i=1}^{n} (\mathbf{X}_{i,k} - \mu_{\mathbf{X}_{*,k}})^2 = \mathrm{var}(\mathbf{X}_{*,k}) .$$

Thus, the variance $\mathrm{var}(\mathbf{X}_{*,k})$ of the k-th data attribute $\mathbf{X}_{*,k}$ can be turned from a double sum expression into the usual mean-centered sum of squares using the pre-computed average $\mu_{\mathbf{X}_{*,k}} = \sum_{i=1}^{n} x_k^i / n$ over the k-th components of all n data items. Since the derivative of squared Euclidean distance yields a decomposition into individual attributes, variance can be computed independently for each data attribute. According to the increased complexity of the derivative of Pearson correlation given in Eqn. 2, plugged into Eqn. 4, such decomposition is not available for Pearson correlation, hence, data components influence each other.

In practice, the derivation for component rating in Pearson correlation can be used like the classical variance, that is, the larger the values of $V_k(\mathbf{X})$ the higher the variability of the k-th attribute in the data set $\mathbf{X}$, see [28], for example. As indicated earlier, only singular values of $\partial r_\lambda(\mathbf{x}, \boldsymbol{w}) / \partial \lambda_k$ lead to problems. Since the current formulation of V_k has got a complexity of $\mathcal{O}(n^2)$ the standard formula of classical variance is preferred in the Euclidean case. A similarly simplified version is not yet known for Pearson correlation.

3.2 Supervised Attribute Rating (SARDUX)

Simultaneous attribute characterization is realized by supervised adaptation of a parametric similarity measure: optimized parameters should minimize dissimilarities within classes while maximizing distances between classes.

Using $\mathrm{d}_\lambda^{ij} = \mathrm{d}_\lambda(\mathbf{x}^i, \mathbf{x}^j)$ for parametrized dissimilarities between n data vectors $\mathbf{x}^i \in \mathbb{R}^d$ $(i = 1 \dots n)$, two double sums σ_1^2, σ_2^2 of pairwise distances are defined:

$$\sigma_l^2(\boldsymbol{\lambda}) = \sum_{i=1}^{n} \sum_{j=1}^{n} \mathrm{d}_\lambda^{ij} \cdot \left(1 - l - (-1)^l \cdot \delta^{ij} \right) , \quad l = 0, 1 . \tag{5}$$

Thereby, σ_1^2 represents data of the same class and σ_0^2 for data of different classes. The Kronecker symbol δ^{ij} indicates identity of class memberships, i.e. $\delta^{ij} = 1$ if the class of data vector $\mathbf{x}^i$ equals the class of vector $\mathbf{x}^j$, $\delta_{ij} = 0$ in case of disagreement. In a one-class scenario, corresponding to the calculation of only σ_1^2 for unlabeled data, the squared Euclidean distance d^2 over all pairs in Eqn. 5 leads to Eqn. 4, i.e. to the calculation of a value proportional to the variance.

Similar to the LDA criterion, a ratio of within-class and between-class distances is used as target of parameter optimization [35]:

$$s = \frac{\sigma_1^2(\boldsymbol{\lambda})}{\sigma_0^2(\boldsymbol{\lambda})} = \min. \tag{6}$$

In the following, the determination of the parameter vector $\boldsymbol{\lambda}$, connected to the adaptive Pearson correlation is of interest [29]. Using the chain rule for the derivatives of the general stress function $\partial s / \partial \lambda_k$, the k-th parameter of the similarity measure can be found by gradient descent with step size γ according to

$$\lambda_k \leftarrow \lambda_k - \gamma \cdot \frac{\partial s}{\partial \lambda_k} \quad \text{with} \quad \frac{\partial s}{\partial \lambda_k} = \sum_{i=1}^{n} \sum_{j=1}^{n} \frac{\partial s}{\partial d_{\boldsymbol{\lambda}}^{ij}} \cdot \frac{\partial d_{\boldsymbol{\lambda}}^{ij}}{\partial \lambda_k}. \tag{7}$$

A locally optimal solution is found for $\frac{\partial s}{\partial \lambda_k} = 0, k = 1 \ldots d$. By applying the quotient rule to the fraction in Eqn. 6 this yields

$$\frac{\partial s}{\partial d_{\boldsymbol{\lambda}}^{ij}} = \frac{\delta_{ij}}{\sigma_0^2} + \frac{(\delta_{ij} - 1) \cdot \sigma_1^2}{(\sigma_0^2)^2}. \tag{8}$$

If classes of patterns i and j are the same, the left term contributes exclusively to the sum, otherwise the right term. The values of σ_0^2 and σ_1^2 must be computed before the calculation of the gradient.

The very right factor $\partial d_{\boldsymbol{\lambda}}^{ij} / \partial \lambda_k$ in Eqn. 7 is already known, because for Pearson correlation, $d_{\boldsymbol{\lambda}}^{ij} = r_{\boldsymbol{\lambda}}^{ij}$, Eqn. 3 can be used with $\mathbf{x} = \mathbf{x}^i$ and $\boldsymbol{w} = \mathbf{x}^j$.

Gradient descent on the target function s can be easily realized by iterative application of Eqn. 7. Typical values of the step size γ are between 0.001 and 1. The gradient can be reused until s increases; then, a recalculation of the gradient is triggered. Optimization can be stopped when no major decrease of s is observable. Depending on the data class distribution it might be advisable to stop after a small number of iterations. This allows to extract a ranked list of variables, while avoiding a potential collapse to the best single remaining data attribute.

It is most natural to initialize $\boldsymbol{\lambda}$ as unweighted Pearson correlation by the unit parameter vector $\lambda_i = 1, i = 1 \ldots d$. After optimization, each vector $\mathbf{x}$ in the data set can be rescaled by its entrywise product with $\boldsymbol{\lambda}$ in order to create a transformed data set for further use with standard correlation methods.

4   Vector Quantization – Neural Gas with Correlation Measure (NG-C)

Neural gas (NG) is a simple, yet, very flexible and powerful centroid-based clustering method [24]. The goal of neural gas is to disperse a number of m data prototypes $\boldsymbol{w}^i \in \mathbf{W}$ among the n data vectors $\mathbf{x}^j$ in such a way that the prototypes represent the data vectors with minimum total quantization error:

$$E(\mathbf{W}, \sigma) = \frac{1}{C(\sigma)} \cdot \sum_{i=1}^{m} \sum_{j=1}^{n} h_\sigma(\mathbf{x}^j, \boldsymbol{w}^i) \cdot d(\mathbf{x}^j, \boldsymbol{w}^i). \tag{9}$$

This sum involves dissimilarities d between data and their prototype vectors. Correlation similarity is thus integrated by the simple expression $d = (1 - r_\lambda)$. The constant $C(\sigma) = \sum_{k=0}^{m-1} h_\sigma(k)$ yields normalization. A combination of a σ-controlled neighborhood range h_σ and a centroid ranking function rnk with

$$h_\sigma(\mathbf{x}^j, \mathbf{w}^i) = \exp(-\mathrm{rnk}(\mathbf{x}^j, \mathbf{w}^i)/\sigma)\,, \quad \mathrm{rnk}(\mathbf{x}^j, \mathbf{w}^i) = \left| \{d(\mathbf{x}^j, \mathbf{w}^k) < d(\mathbf{x}^j, \mathbf{w}^i)\} \right|$$

leads to a competitive learning scheme with neighborhood cooperation. Rank operations are useful for breaking ties in similarity relationships frequently occurring in high-dimensional data. Prototype adaptation is realized by a stochastic gradient descent on the error function given in Eqn. 9. The optimal direction for a considered prototype $\mathbf{w}^j$ given a presented pattern $\mathbf{x}^j$ is obtained as derivative $\partial E(\mathbf{x}^j, \sigma)/\partial \mathbf{w}^i$. This is based on $\partial d(\mathbf{x}^j, \mathbf{w}^i)/\partial w_k^i$ which for the Pearson correlation dissimilarity $d = (1 - r_\lambda)$ is expressed by the negative of Eqn. 2, leading to gradient ascent. In summary, a very simple clustering algorithm is obtained [33]:

Fix a number m of prototypes.
Initialize all prototypes $\mathbf{w}^i, i = 1 \ldots m$ by randomly drawn data vectors.
Repeat: Randomly draw $\mathbf{x}^j$.
For all centroids $\mathbf{w}^i$ and all vector components k:

$$w_k^i \leftarrow w_k^i + \gamma \cdot h_\sigma(\mathbf{x}^j, \mathbf{w}^i) \cdot \frac{\partial r_\lambda(\mathbf{x}, \mathbf{w})}{\partial w_k}\,.$$

This correlation-based neural gas algorithm (NG-C) drives the centroids in steps of γ iteratively to optimal locations. Since derivatives of correlation might be very close to zero, reasonable learning rates lie in large intervals, such as $\gamma \in [0.001; 10000]$. Usually, good convergence is reached after 100–1000 repeated data cycles, depending on the number of data vectors. During training, the neighborhood range σ is exponentially decreased from a starting size of $\sigma = m$ to a small value like $\sigma = 0.001$. This involves all prototypes strongly in the beginning, contracting centroids towards the data 'center', and it leads to a fine-tuning of data-specific centroids in the final training phase. Since the neighborhood size is changing rather slowly in contrast to the relatively fast centroid update, the neighborhood-related normalization term $1/C(\sigma)$ in Eqn. 9 can be omitted in the iteration loop. For easier interpretation of the data prototypes, the parameter vector λ is fixed to $\lambda = 1$, although simultaneous adaptation of a global or even of prototype-specific parameters λ is possible, in principle.

As a reminder, the k-means clustering algorithm can also be used for clustering with correlation measure, if a z-score transform is applied to the data vectors. In that case, the partitions from NG-C and k-means are theoretically identical. However, many packages, among which the widely used Cluster 3.0 package, do not allow a z-score transform as preprocessing step. In these cases NG-C provides better quantization results.

A simple example of quantization in correlation space is given for the three data vectors $\mathbf{x}^1 = (3, 0, 0)$, $\mathbf{x}^2 = (2, 3, 0)$ and $\mathbf{x}^3 = (0, 2, 1)$. For their mean vector $\mu = (1.67, 1.67, 0.33)$ an average correlation $\bar{r}_\mu = 0.482 \pm 0.473$ is obtained, while for a correlation-optimized vector $\mathbf{w} = (0.7, 1, 0)$ the result can be improved to $\bar{r}_\mu = 0.506 \pm 0.429$. Moreover, the concept of NG-C remains valid for other differentiable dissimilarity measures, which does not hold for k-means in general.

For classification tasks, modifications of the prototype-based learning vector quantization (LVQ) are available to obtain simultaneous update of prototypes and the parameters of an adaptive correlation measure [30]. Classification quality can be much simpler assessed than the quality of clustering. The former allows to compute classification errors or receiver operator characteristic (ROC) curves independently of the model, while in the latter case prototype-specific clustering measures, such as the quantization error, depend on the number of prototypes, for example. In practice it is difficult to define the trade-off between data compression, data quantization accuracy, and extracted cluster shapes. This might lead to the perception of clustering as an ill-posed, application-dependent problem.

NG-C is available at `http://pgrc-16.ipk-gatersleben.de/~stricker/ng/` as C source code.

5   Visualization – High-Throughput Multidimensional Scaling (HiT-MDS)

Many domains of data analysis require a faithful reduction of data dimensionality. Apart from manual attribute picking, the most commonly used tool for the compression of high-dimensional data is the projection to principal components as obtained by PCA. It is a fast linear projection method in favor of the directions of maximum variance in the data space. A more intuitive approach to faithful dimension reduction aims at proper reconstruction of original data distances in a low-dimensional scatter plot. Multidimensional scaling (MDS) and other nonlinear dimension reduction techniques can be used for that purpose [7,17,36]. In the following HiT-MDS, a method for embedding high-throughput data, is revisited. It maximizes the correlation between the original data distance matrix and the distance matrix of a reconstruction space with adaptive surrogate points [32]. This is a strong benefit over traditional methods where distances of the surrogate points are optimized in least-squares relation to the original distances, thereby imposing unnecessary constraints on the optimization of the embedding stress function [9].

An optimization task is devised to find locations of points $\hat{x}^i = (\hat{x}^i_1, \ldots, \hat{x}^i_q)$, $\hat{x}^i \in \hat{X}_{n \times q}$ in a q-dimensional target space corresponding to the $i = 1 \ldots n$ input vectors $x^i \in X_{n \times d}$ of dimension d. Let matrix $D = (d^{ij})_{i,j=1\ldots n}$ contain pattern distances, and $\hat{D} = (\hat{d}^{ij})_{i,j=1\ldots n}$ those of the reconstructions. HiT-MDS maximizes the correlation between entries of the source distance matrix D and the reconstructed distances $\hat{D}$ by minimizing the embedding cost function s:

$$s = -r_\lambda \left(D, \hat{D}(\hat{X}) \right) = \min \quad \text{with} \quad \frac{\partial s}{\partial \hat{x}^i_k} = - \sum_{\substack{j=1\ldots n}}^{j \neq i} \frac{\partial r_\lambda}{\partial \hat{d}^{ij}} \cdot \frac{\partial \hat{d}^{ij}}{\partial \hat{x}^i_k}, i = 1 \ldots n \quad (10)$$

Locations of points in target space are driven by stochastic gradient descent on the stress function s. Randomly drawn points $\hat{\mathbf{x}}^i$ are updated, for rapid convergence, into the direction of the sign $\text{sgn}(x) = x/|x|$ of the steepest gradient of s:

$$\hat{x}_k^i \leftarrow \hat{x}_k^i - \gamma \cdot \text{sgn}(\frac{\partial s}{\partial \hat{x}_k^i}) . \qquad (11)$$

For Euclidean output spaces the rightmost factor in Eqn. 10 is

$$\frac{\partial \hat{d}^{ij}}{\partial \hat{x}_k^i} = (\hat{x}_k^i - \hat{x}_k^j)/\hat{d}^{ij} \quad \text{for distance} \quad \hat{d}^{ij} = \sqrt{\sum_{l=1}^{d}(\hat{x}_l^i - \hat{x}_l^j)^2} \qquad (12)$$

The correlation derivative $\frac{\partial r_\lambda}{\partial \hat{d}^{ij}}$ required in Eqn. 10 is given as back-reference to Eqn. 2 for vectorized matrices $\mathbf{x} = \mathbf{D}$ and $\mathbf{w} = \hat{\mathbf{D}}$. After updating a single point $\hat{\mathbf{x}}^i$ in relation to the other points it is not necessary to recompute the stress function s in Eqn. 10 for the whole matrix $\hat{\mathbf{D}}(\hat{\mathbf{X}})$, because only its i-th row and column have changed. For that case an incremental update of s is available [34].

The only parameters of HiT-MDS are the learning rate γ, which can be quite robustly chosen from the interval $\gamma \in [0.01; 100]$, and a stopping criterion like the number of iterations. In order to enhance the early update phase, surrogate points $\hat{\mathbf{x}}^i$ can be initialized by random projection of the original data into the target space, which is better than completely random initialization [14]. Several runs of HiT-MDS might be considered to resolve ambiguities for point placement in the late update phase, leading to locally optimal solutions. Standardized orientations of the final low-dimensional embeddings can be cheaply obtained by PCA post-processing, which is helpful, because the stress function s is invariant to scaling, rotation, and translation of point configurations in the surrogate space. Different embeddings can be compared by Procrustes analysis [22], the quality of an embedding in relation to the original data by the concept of trustworthiness and more general rank-based neighborhood criteria [15,18].

As usual in MDS approaches, data need not be given in form of high-dimensional vectors – a distance matrix $\mathbf{D}$ is sufficient. Then, surrogate points in the target space can be initialized randomly or, if time permits, by the linear MDSLocalize approach [31]. Virtually any type of data dissimilarity matrix $\mathbf{D}$ can be used for embedding with HiT-MDS, however, it might be impossible to reach satisfying results in the Euclidean target space realized by Eqn. 12.

C, MATLAB (GNU Octave), and R source code of high-throughput multidimensional scaling is online available at `http://hitmds.webhop.net/`.

6   Analysis of Biological Data

In biomedical studies of the tissue transcriptome in plants and animals, array-based cDNA hybridization is one of the most frequently used technologies for measuring thousands of gene activities in parallel. In the presented case, the transcript of barley seed development is examined at 14 equidistant time points of two days, at days 0–26 after flowering. Macroarrays were used for extracting

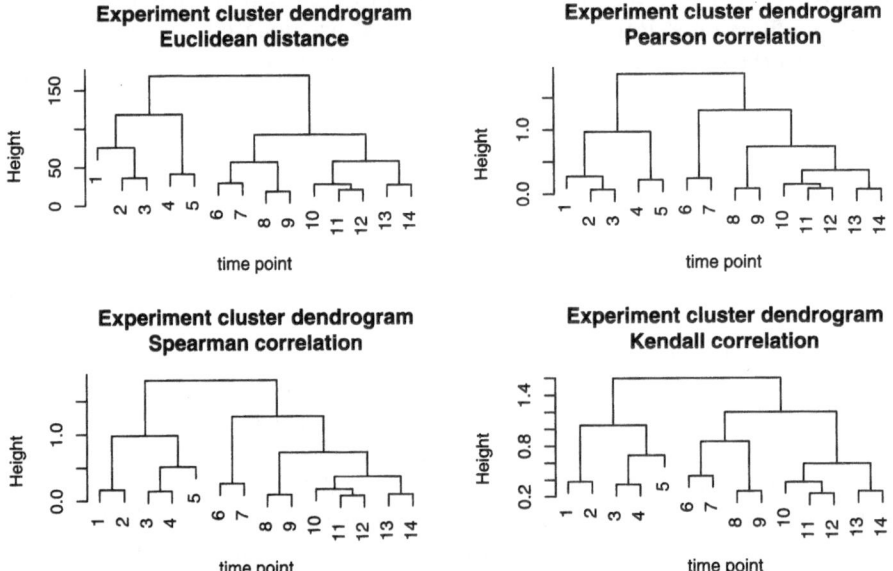

Fig. 2. Hierarchical clusterings of normalized gene expression experiments using complete linkage and optimum leaf ordering

expression levels describing the regulation status of 11,786 genes per time point. Using two biological and two technical replicates, filter criteria led to a subset of 4,824 high quality genes of which the expression levels were normalized and $\log_2$-transformed [32]. Questions regarding the data concern the general data characterization and the identification of genes active at specific temporal stages.

6.1   Experiment Clustering

Data validity can be tested by checking the temporal integrity of the data. Since non-cyclic tissue development is considered, a directed arrangement of the experiments can be expected. Their similarity-based grouping can be obtained by hierarchical clustering [13]. The resulting dendrograms are not unique, because subtrees can be freely rotated around nodes, therefore, an optimum leaf ordering strategy was applied [4] in order to obtain the unique clustering results displayed in Fig. 2. Four different data measures, Euclidean distance, Pearson, Spearman and Kendall correlation, were used for complete linkage hierarchical clustering. Remarkably, all measures reconstruct orderings in perfect agreement with the experimental setup which is a good indication of reliable data quality. Considering this global level of comparison, involving 4,824 genes per experiment, only minor differences can be found in the dendrograms. Particularly, the gap between the two time intervals 1–5 and 6–14 is well-conserved, regardless of the measure. This gap denotes an intermediate stage of strong transcriptional reprogramming happening between early seed development and the late nutrient storage phase.

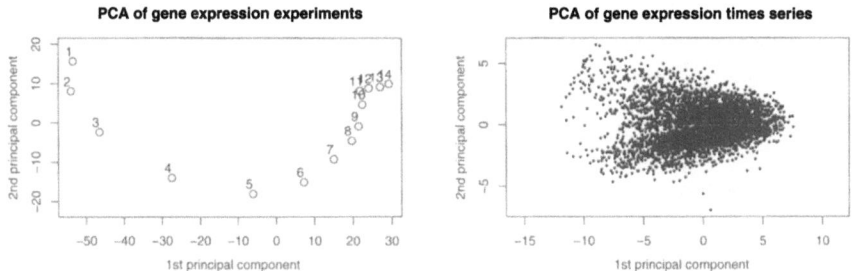

Fig. 3. Principal component projections. Left: 14 gene experiments containing expression levels for 4,824 genes. Right: 4,824 genes as expression time series of length 14.

On a local scale of comparison, though, quite large differences become apparent for the contribution of specific genes to the underlying comparison measure. This is demonstrated for the variance and its correlation-based counterpart later. The observed directed experiment grouping is essentially caused by temporal development that affects many genes simultaneously.

6.2   Principal Component Plots of Experiments and Genes

Another global perspective on the data is obtained by the projection to the first two principal components which explain 60.4% of the overall variance. Like the dendrograms, the resulting scatter plot in the left panel of Fig. 3 shows a consistent temporal ordering of the experiments. A fast transition is found at time points 3 to 6, and a stagnation of development in the late stages 6–14 is indicated by a progressive point densification. The assessment of the inter-gene relationships of the 4,824 expression time series by principal component projection yields the non-informative scatter plot displayed in the right panel of Fig. 3 to be complemented by the correlation-based visualization result of HiT-MDS in Fig. 7, discussed in detail below.

6.3   Unsupervised Attribute Rating

The identification of attributes that strongly contribute to the overall relationships among data points helps to characterize the data set as a whole. In the Euclidean case the deviation of attributes from the common center of gravity can be assessed – this is the variance. The situation is different for Pearson correlation, because there is not only one such center, but infinitely many optimal vectors with maximum correlation to the data. The attribute rating is obtained by Eqn. 4. The characterization of the 14 time points in the 4,824 gene expression series is summarized in Fig. 4. Variance shown on the left decreases to a minimum at time point 4, followed by an intermediate variance until time point 10, and an increase again in the late stage. This is quite different from the correlation-based attribute rating, where an intermediate increase is completely missing.

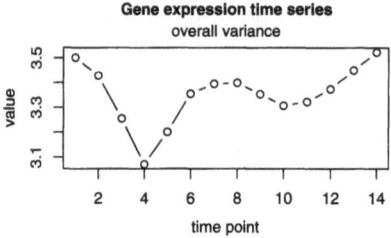 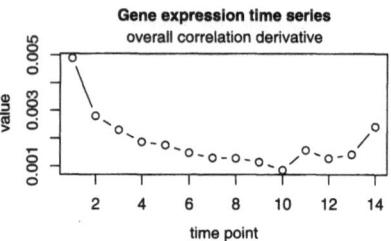

Fig. 4. Unsupervised characterization of time points in developing barley grain transcriptome. Left: ordinary statistical variance. Right: variance analogon for Pearson correlation calculated by Eqn. 4.

In other words, the correlation of experimental expression patterns changes gradually, while the absolute gene expression values, measured by Euclidean distance, change rapidly at time point 4. Such a complementary assessment is very interesting tool for any study related to the analysis of changes in expression patterns in general.

6.4 Supervised Attribute Rating

Auxiliary information allows further qualification of the data attributes. Since the interest refers to genes involved in temporal seed development, a canonic binary categorization into early and late stage is considered next. Due to known transcriptional reprogramming, supported by the dendrograms in Fig. 2, important processes occur between early and late stages. This requires additional categories for a better characterization of this intermediate phase. Four groups were created for time intervals 1–2, 3–5, 6–9, and 10–14, according to biological background knowledge. The list of genes ranked after SARDUX training contains a strict temporal organization of the underlying time series. This is illustrated for the top 25 genes – corresponding to the 25 largest components in λ – for which the gene expression profiles are plotted in the left panel of Fig. 5. Up-regulation events are more prominent during development than down-regulation which is in accordance with the array design. This design is specific to development-related genes in cereals, which induces a visible bias towards increasing and decreasing time series of genetic under- and over-expression. The top rated genes trigger cell extension and cell wall degradation which are typical features of seed development. A contrasting sample of 25 randomly chosen genes is displayed in the right panel where a consistent functional structure is missing. After all, genes capturing the intrinsic temporal processes of the experiments were identified by SARDUX.

6.5 Correlation-Based Browser of Expression Patterns

Distance-preserving plots are indispensable tools for visual data inspection. The data projection to the first two principal components is a standard approach to

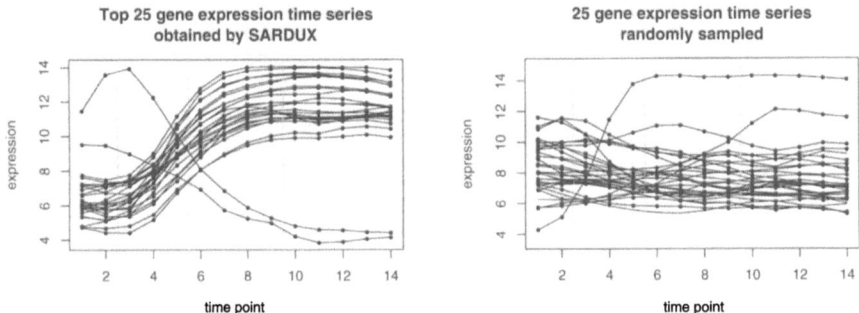

Fig. 5. Top 25 gene expression time series obtained by correlation-based SARDUX relevance rating are shown in the left panel. For comparison, 25 randomly selected genes are displayed on the right.

reach this aim. Its limitation for the visualization of gene expression time series was indicated in a previous section, though. Kohonen's self-organizing maps provide a superior visualization by non-linear, density dependent mappings of data to a grid of vector quantizing nodes [15]. However, such mappings lead to a loss of the individuality of data points. Multidimensional scaling approaches maintain the correspondence of data points to their embedded counterparts while aiming at preserving distance relationships. Planar scatter plots of high-dimensional experiments and short gene expression time series are given in this section.

Experiment view. The left panel in Fig. 6 contains the resulting scatter plot of HiT-MDS for a correlation-based view on the gene expression experiments. In contrast to the linear PCA projection in Fig. 3, the MDS scatter plot does not exhibit the bellied time trajectory, and it shows more details in the very early and late developmental stages. The symbols denote stages used for the SARDUX-based rating of genes as given in the previous section. The effect of SARDUX applied to the 4,824-dimensional labeled data is shown in the right panel of Fig. 6. The restructuring of embedded experiments results from genes for which between-class separation and within-class shrinkage get emphasized. In the HiT-MDS visualization, points with different class labels get better separated, particularly between time points 5 and 6.

Gene view. Complementary to the visual assessment of experiment data, a browsable scatter plot of gene expression time series helps to identify major regulatory modes and general regulation patterns. The within separation of groups of very similarly up- and down-regulated gene expression time series can be much enhanced when the dissimilarity measure $(1 - r)$ is replaced by its higher-order power $(1 - r)^p$ [41]. For the data at hand, a choice of $(1 - r)^8$ is considered a good dissimilarity measure for the visual separation of the 4,824 gene expression time series [32]. The corresponding display of embedded gene expression time series is shown in either panel of Fig. 7, disregarding point sizes for the moment.

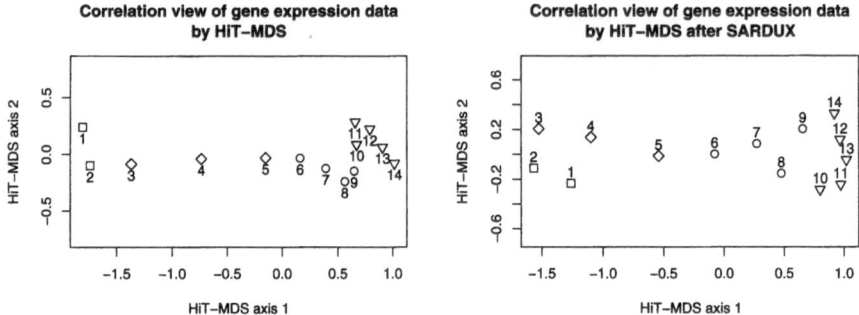

Fig. 6. HiT-MDS embeddings of Pearson correlation $(1 - r)$ between gene expression experiments in the Euclidean plane. Different symbol labels denote categories of crucial developmental stages in barley grains. Left: nonlinear alternative to the principal component projection as given in the right panel of Fig. 3. Right: same data with attributes rescaled by SARDUX, using the four class categories.

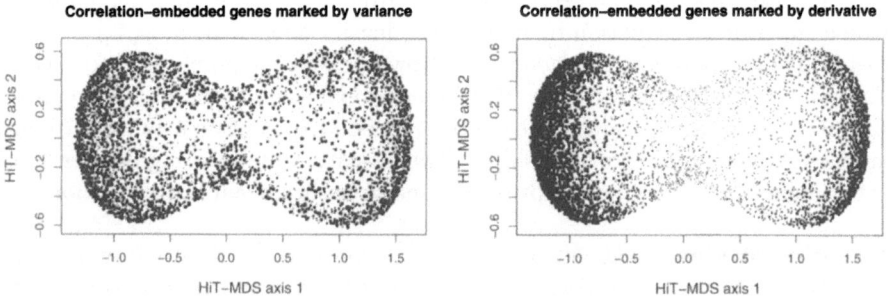

Fig. 7. HiT-MDS embeddings of 4,824 gene expression time series in Euclidean space. Powers of Pearson correlation $(1 - r)^8$ are used as similarity measure. Left: Point sizes are proportional to the statistical variance of the genes. Right: point sizes represent values of the correlation variability given by Eqn. 4. The plots contain point configurations alternative to the PCA visualization in the right panel of Fig. 3: the left balloon parts correspond to down-regulated time series, the right ones to up-regulated series, and the waist parts to intermediate gene up- and down regulation, as detailed in Fig. 8.

A clear difference of this lying sand-glass shape to the PCA projection in the right panel of Fig. 3 can be stated. The most prominent regulatory patterns associated with points at the top and bottom waist regions and with points at the left and right of the balloon-shaped parts are displayed in Fig. 8 for local subsets of the data obtained by NG-C vector quantization as discussed in the next section. The complementary view of Fig. 7 and Fig. 8 demonstrates a good association of nearby points in MDS and the underlying profiles of expression regulation. After all, substantial improvement over the less structured PCA plot in the right panel of Fig. 3 is obtained.

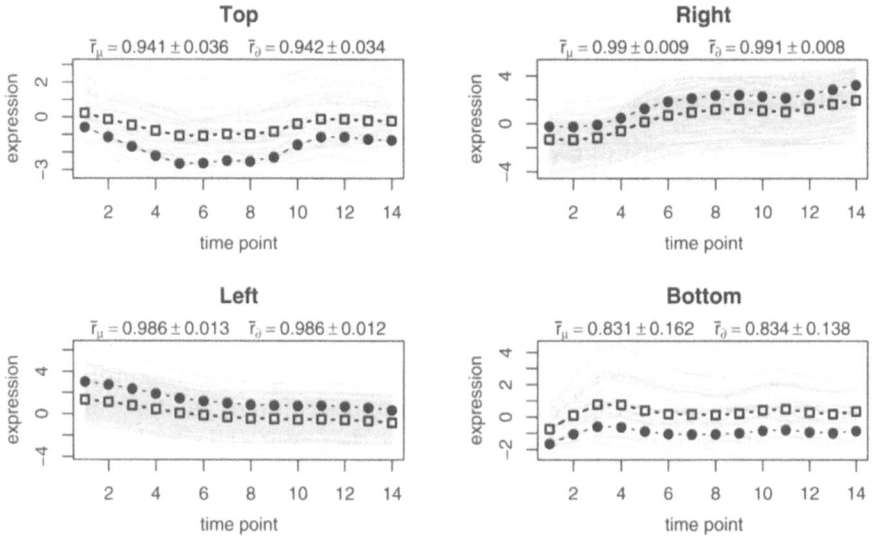

Fig. 8. Gene expression time series associated with four extreme locations of Fig. 7. The average lines carry box symbols, the correlation-based NG-C centers are decorated by bullets. NG-C centers are invariant to vertical offset and scaling, which can only be visualized by specifically shown examples. Average correlations between mean ($\bar{r}_\mu$) and NG-C centers ($\bar{r}_\partial$) are given on top for comparison.

Point sizes underline the adequacy of the MDS displays. The bigger a point, the more outstanding the corresponding gene in under a measure-specific view. The left panel contains point sizes proportional to the variance of the expression time series, whereas the right panel contains point sizes proportional to the correlation-based derivative given by Eqn. 2. A clear incompatibility of variance and correlation-based analysis is indicated by the lack of a structured distribution of point sizes in the left panel. Conversely, the smooth gradients from correlation-based gene ratings according to Eqn. 2 in the right panel support a good match of the correlation-based MDS display and the unsupervised gene rating: genes in the inside are very conformant, while extremal genes on the data boundary produce high tension by their strong variability.

6.6   Neural Gas Clustering

The analysis of the gene expression data set is closed by clustering the 4,824 expression time series into $k = 7$ major partitions. For that purpose NG-C and k-means were applied 25 times for assessing the average correlation of all prototypes with data in their receptive fields. The baseline clustering is k-means with Euclidean distance, as provided in many standard bioinformatics tools. This is complemented by k-means applied to z-score transformed data and to correlation-optimized NG-C. According to the choice of correlation similarity, results compiled in Tab. 2 point out the benefits of Pearson correlation for the

Table 2. Average correlations and standard deviations over 25 runs of 7 k-means and NG-C prototypes with data in their receptive fields of 4,824 gene expression time series. Non-significant differences between k-means (correlation) and NG-C result from algorithm-specific termination criteria.

k-means (Euclidean)	**k-means** (Pearson correlation)	**NG-C**
0.87559 ± 0.00158	0.91958 ± 0.00030	0.91976 ± 0.00002

quantization accuracy. However, no clear difference can be stated between k-means with z-score transformed data and NG-C. This is in agreement with the fact that, theoretically, both approaches should converge to the same optima.

A localized view reveals the main temporal structure occurring in the set of gene expression time series. Two major structures of up-regulation and down-regulation are prominent, corresponding to the outer right and left balloon shapes in Fig 7. These are linked by genes of intermediate up- and down-regulation. Representative sequence clusters from these four opposing dynamics are shown in Fig. 8. Also shown are the underlying NG-C prototypes together with the averages, trivially equivalent to Euclidean 1-mean(s) clustering, of all sequences in the depicted clusters. Since the information transfer between data and prototype representation is not controlled in this experiment, the prototype densities are not equal to the data densities; this explains, why clusters contain quite different numbers of sequences. If desired the information transfer could be regulated by magnification control [37] which is one principal strength of NG methods, but the results would not be comparable to k-means in that case.

For completeness, average correlations and standard deviations of data sequences with their means and their NG-C prototypes are reported above each image panel. Only very small differences of the quantization accuracies in favor of NG-C can be found on that very local perspective with already highly correlated expression patterns.

7   Discussion

Derivatives of Pearson correlation offer consistent approaches to optimized models of data representation. Pearson correlation is widely used in biomedical applications and allows a proper mathematical implementation in gradient-based frameworks. As argued in the section on correlation measures, some functionality presented in this article could be simulated by the squared Euclidean distance via a z-score transform of the data vectors. This is especially the case for prototype-based clustering, where k-means for transformed data produced equally good results like the NG-C clustering method. The concept realized by NG is yet more general and suitable for any other differentiable data measure. Moreover, NG-C is an online method, whereas k-means runs in batch mode. Yet, it would be interesting to add correlation to the rapidly converging batch NG [8] combined with the concept of magnification for controlling the information transfer between data and prototypes [37].

The unsupervised and supervised attribute rating schemes profit a lot from the presented derivatives of correlation. The analogon of variance offers new perspectives on pattern-based variability of data attributes, whereas the supervised SARDUX rating allows to detect class-specific variable contribution, useful in tasks of biomarker design.

In case of multidimensional scaling, the optimization of the Pearson correlation between the original data similarity matrix and the reconstructed matrix offers many more degrees of freedom for the optimized space reconstruction than the traditional least squares minimization between the distances. Faithful correspondence between the low-dimensional reconstruction and the original space can be obtained, which has been demonstrated for gene expression time series compared by the $8th$ power of correlation. The computational demands of HiT-MDS are especially challenged by the limited memory storage capacity on 32 bit desktop computers and by potentially long processing times of huge distance matrices. The efforts are compensated by great versatility, because arbitrary and sparse similarity matrices can be reconstructed very fast in comparison to other available MDS schemes.

8   Conclusions

A uniform approach based on cost function optimization has been presented for correlation-based data analysis. Thereby, derivatives of a parametric extension of the Pearson correlation coefficient are used to address three major aspects of analysis: attribute characterization, data clustering and visualization that can be smoothly interlinked for creating comprehensive views of the data. The methods were applied to a single gene expression data set for providing an illustrative of their utility; further examples can be found in the references. The described approaches can be used in other domains of data processing too. Finally, the concepts can be easily generalized by replacing the derivative of Pearson correlation by that of any other suitable similarity measure for which a parametric formulation and the derivatives are available.

Acknowledgment

The work is supported by grant XP3624HP/0606T, Ministry of Culture Saxony-Anhalt.

References

1. Anscombe, F.J.: Graphs in statistical analysis. American Statistician 27, 17–21 (1973)
2. Azuaje, F., Dopazo, J.: Data Analysis and Visualization in Genomics and Proteomics. Wiley, Chichester (2005)
3. Balasubramaniyan, R., Hüllermeier, E., Weskamp, N., Kämper, J.: Clustering of gene expression data using a local shape-based similarity measure. Bioinformatics 21(7), 1069–1077 (2005)

4. Bar-Joseph, Z., Gifford, D.K., Jaakkola, T.S.: Fast optimal leaf ordering for hierarchical clustering. Bioinformatics 17(suppl. 1), S22–S29 (2001)
5. Blest, D.: Rank correlation – an alternative measure. Australian & New Zealand Journal of Statistics 42(1), 101–111 (2000)
6. Bloom, J., Adami, C.: Apparent dependence of protein evolutionary rate on number of interactions is linked to biases in protein-protein interactions data sets. BMC Evolutionary Biology 3(1), 21 (2003)
7. Buja, A., Swayne, D., Littman, M., Dean, N., Hofmann, H.: Interactive Data Visualization with Multidimensional Scaling. Report, University of Pennsylvania (2004), http://www-stat.wharton.upenn.edu/~buja/
8. Cottrell, M., Hammer, B., Hasenfuß, A., Villmann, T.: Batch NG. In: Verleysen, M. (ed.) European Symposium on Artificial Neural Networks (ESANN), pp. 275–282. D-side Publications (2005)
9. Cox, M., Cox, M.: Multidimensional Scaling. Chapman and Hall, Boca Raton (2001)
10. Ferguson, T., Genest, C., Hallin, M.: Kendall's Tau for autocorrelation. The Canadian Journal of Statistics 28(3), 587–604 (2000)
11. Gersho, A., Gray, R.M.: Vector Quantization and Signal Compression. Springer, Heidelberg (1992)
12. Hartigan, J.A., Wong, M.A.: A K-means clustering algorithm. Applied Statistics 28, 100–108 (1979)
13. Johnson, S.: Hierarchical Clustering Schemes. Psychometrika 2, 241–254 (1967)
14. Kaski, S.: Dimensionality reduction by random mapping: Fast similarity computation for clustering. In: Proceedings of the International Joint Conference on Neural Networks (IJCNN 1998), vol. 1, pp. 413–418. IEEE Service Center, Piscataway (1998)
15. Kaski, S., Nikkila, J., Oja, M., Venna, J., Toronen, P., Castren, E.: Trustworthiness and metrics in visualizing similarity of gene expression. BMC Bioinformatics 4(1), 48 (2003)
16. Kohonen, T.: Self-Organizing Maps, 3rd edn. Springer, Berlin (2001)
17. Lee, J., Verleysen, M.: Nonlinear Dimension Reduction. Springer, Heidelberg (2007)
18. Lee, J., Verleysen, M.: Rank-based quality assessment of nonlinear dimensionality reduction. In: Verleysen, M. (ed.) European Symposium on Artificial Neural Networks (ESANN), pp. 49–54. D-facto Publications (2008)
19. Lodhi, H., Saunders, C., Shawe-Taylor, J., Cristianini, N., Watkins, C.: Text classification using string kernels. Journal of Machine Learning Research 2, 419–444 (2002)
20. Lohninger, H.: Teach/Me Data Analysis. Springer, Heidelberg (1999)
21. Ma, Y., Lao, S., Takikawa, E., Kawade, M.: Discriminant analysis in correlation similarity measure space. In: Ghahramani, Z. (ed.) Proceedings of the 24th Annual International Conference on Machine Learning (ICML 2007), pp. 577–584. Omnipress (2007)
22. Mardia, K., Dryden, I.: Statistical Shape Analysis. Wiley, Chichester (1998)
23. Martinetz, T., Berkovich, S., Schulten, K.: "Neural-gas" network for vector quantization and its application to time-series prediction. IEEE Transactions on Neural Networks 4(4), 558–569 (1993)
24. Martinetz, T., Schulten, K.: A "neural-gas" network learns topologies. Artificial Neural Networks I, 397–402 (1991)
25. Meuleman, W., Engwegen, J., Gast, M.-C., Beijnen, J., Reinders, M., Wessels, L.: Comparison of normalisation methods for surface-enhanced laser desorption and

ionisation (SELDI) time-of-flight (TOF) mass spectrometry data. BMC Bioinformatics 9(1), 88 (2008)

26. Nielsen, N., Carstensen, J., Smedsgaard, J.: Aligning of single and multiple wavelength chromatographic profiles for chemometric data analysis using correlation optimised warping. Journal of Chromatography 805, 17–35 (1998)

27. Sreenivasulu, N., Radchuk, V., Strickert, M., Miersch, O., Weschke, W., Wobus, U.: Gene expression patterns reveal tissue-specific signaling networks controlling programmed cell death and ABA-regulated maturation in developing barley seeds. The Plant Journal 47(2), 310–327 (2006)

28. Strickert, M., Schleif, F.-M., Seiffert, U., Villmann, T.: Derivatives of Pearson correlation for gradient-based analysis of biomedical data. Inteligencia Artificial, Revista Iberoamericana de IA 12(37), 37–44 (2008)

29. Strickert, M., Schleif, F.-M., Villmann, T.: Metric adaptation for supervised attribute rating. In: Verleysen, M. (ed.) European Symposium on Artificial Neural Networks (ESANN), pp. 31–36. D-facto Publications (2008)

30. Strickert, M., Seiffert, U., Sreenivasulu, N., Weschke, W., Villmann, T., Hammer, B.: Generalized relevance LVQ (GRLVQ) with correlation measures for gene expression data. Neurocomputing 69, 651–659 (2006)

31. Strickert, M., Sreenivasulu, N., Seiffert, U.: Sanger-driven MDSLocalize - A comparative study for genomic data. In: Verleysen, M. (ed.) European Symposium on Artificial Neural Networks (ESANN), pp. 265–270. D-facto Publications (2006)

32. Strickert, M., Sreenivasulu, N., Usadel, B., Seiffert, U.: Correlation-maximizing surrogate gene space for visual mining of gene expression patterns in developing barley endosperm tissue. BMC Bioinformatics 8(165) (2007)

33. Strickert, M., Sreenivasulu, N., Villmann, T., Hammer, B.: Robust centroid-based clustering using derivatives of Pearson correlation. In: Proc. Int. Joint Conf. Biomedical Engineering Systems and Technologies, BIOSIGNALS, Madeira (2008)

34. Strickert, M., Teichmann, S., Sreenivasulu, N., Seiffert, U.: High-Throughput Multi-Dimensional Scaling (HiT-MDS) for cDNA-array expression data. In: Duch, W., Kacprzyk, J., Oja, E., Zadrożny, S. (eds.) ICANN 2005. LNCS, vol. 3696, pp. 625–633. Springer, Heidelberg (2005)

35. Strickert, M., Witzel, K., Mock, H.-P., Schleif, F.-M., Villmann, T.: Supervised attribute relevance determination for protein identification in stress experiments. In: Proceedings of Machine Learning in Systems Biology (MLSB 2007), pp. 81–86 (2007)

36. Venna, J., Kaski, S.: Neighborhood preservation in nonlinear projection methods: An experimental study. In: Dorffner, G., Bischof, H., Hornik, K. (eds.) Proceedings of the International Conference on Artificial Neural Networks (ICANN), pp. 485–591. Springer, Heidelberg (2001)

37. Villmann, T., Claussen, J.C.: Magnification control in self-organizing maps and neural gas. Neural Computation 18(2), 446–469 (2006)

38. Villmann, T., Schleif, F.-M., Hammer, B.: Comparison of Relevance Learning Vector Quantization with other Metric Adaptive Classification Methods. Journal of Neural Networks 19(5), 610–622 (2006)

39. Xu, W., Chang, C., Hung, Y., Kwan, S., Fung, P.: Order Statistics Correlation Coefficient as a Novel Association Measurement with Applications to Biosignal Analysis. IEEE Transactions on Signal Processing 55(12), 5552–5563 (2007)

40. Yang, L.: An overview of distance metric learning. Technical report, Department of Computer Science and Engineering, Michigan State University (2007)

41. Zhou, X., Kao, M.-C.J., Wong, W.H.: Transitive functional annotation by shortest-path analysis of gene expression data. PNAS 99(20), 12783–12788 (2002)

Median Topographic Maps
for Biomedical Data Sets

Barbara Hammer[1], Alexander Hasenfuss[1], and Fabrice Rossi[2]

[1] Clausthal University of Technology, D-38678 Clausthal-Zellerfeld, Germany
[2] INRIA Rocquencourt, Domaine de Voluceau, Rocquencourt, B.P. 105,
78153 Le Chesnay Cedex, France

Abstract. Median clustering extends popular neural data analysis
methods such as the self-organizing map or neural gas to general data
structures given by a dissimilarity matrix only. This offers flexible and
robust global data inspection methods which are particularly suited for
a variety of data as occurs in biomedical domains. In this chapter, we
give an overview about median clustering and its properties and extensions, with a particular focus on efficient implementations adapted to
large scale data analysis.

1 Introduction

The tremendous growth of electronic information in biological and medical domains has turned automatic data analysis and data inspection tools towards a
key technology for many application scenarios. Clustering and data visualization
constitute one fundamental problem to arrange data in a way understandable by
humans. In biomedical domains, prototype based methods are particularly well
suited since they represent data in terms of typical values which can be directly
inspected by humans and visualized in the plane if an additional low-dimensional
neighborhood or embedding is present. Popular methodologies include K-means
clustering, the self-organizing map, neural gas, affinity propagation, etc. which
have successfully been applied to various problems in the biomedical domain
such as gene expression analysis, inspection of mass spectrometric data, health-
care, analysis of microarray data, protein sequences, medical image analysis, etc.
[1,36,37,41,44,53,54].

Many popular prototype-based clustering algorithms, however, have been derived for Euclidean data embedded in a real-vector space. In biomedical applications, data are diverse including temporal signals such as EEG and EKG signals,
functional data such as mass spectra, sequential data such as DNA sequences,
complex graph structures such as biological networks, etc. Often, the Euclidean
metric is not appropriate to compare such data, rather, a problem dependent
similarity or dissimilarity measure should be used such as alignment, correlation,
graph distances, functional metrics, or general kernels.

Various extensions of prototype-based methods towards more general data
structures exist such as extensions for recurrent and recursive data structures,

M. Biehl et al.: (Eds.): Similarity-Based Clustering, LNAI 5400, pp. 92–117, 2009.

functional versions, or kernelized formulations, see e.g. [7,24,25,26,27] for an overview. A very general approach relies on a matrix which characterizes the pairwise similarities or dissimilarities of data. This way, any distance measure or kernel (or generalization thereof which might violate symmetry, triangle inequality, or positive definiteness) can be dealt with including discrete settings which cannot be embedded in Euclidean space such as alignment of sequences or empirical measurements of pairwise similarities without explicit underlying metric.

Several approaches extend popular clustering algorithms such as K-means or the self-organizing map towards this setting by means of the relational dual formulation or kernelization of the approaches [8,24,30,31,51]. These methods have the drawback that they partially require specific properties of the dissimilarity matrix (such as positive definiteness), and they represent data in terms of prototypes which are given by (possibly implicit) mixtures of training points, thus they cannot easily be interpreted directly. Another general approach leverages mean field annealing techniques [19,20,33] as a way to optimize a modified criterion that does not rely anymore on the use of prototypes. As for the relational and kernel approaches, the main drawback of those solutions is the reduced interpretability.

An alternative is offered by a representation of classes by the median or centroid, i.e. prototype locations are restricted to the discrete set given by the training data. This way, the distance of data points from prototypes is well-defined. The resulting learning problem is connected to a well-studied optimization problem, the K-median problem: given a set of data points and pairwise dissimilarities, find k points forming centroids and an assignment of the data into k classes such that the average dissimilarities of points to their respective closest centroid is minimized. This problem is NP hard in general unless the dissimilarities have a special form (e.g. tree metrics), and there exist constant factor approximations for specific settings (e.g. metrics) [6,10]. The popular K-medoid clustering extends the batch optimization scheme of K-means to this restricted setting of prototypes: it in turn assigns data points to the respective closest prototypes and determines optimum prototypes for these assignments [9,38]. Unlike K-means, there does not exist a closed form of the optimum prototypes given fixed assignments such that exhaustive search is used. This results in a complexity $\mathcal{O}(N^2)$ for one epoch for K-centers clustering instead of $\mathcal{O}(N)$ for K-means, N being the number of data points. Like K-means, K-centers clustering is highly sensitive to initialization.

Various approaches optimize the cost function of K-means or K-median by different methods to avoid local optima as much as possible, such as Lagrange relaxations of the corresponding integer linear program, vertex substitution heuristics, or affinity propagation [17,18,28]. In the past years, simple, but powerful extensions of neural based clustering to general dissimilarities have been proposed which can be seen as generalizations of K-centers clustering to include neighborhood cooperation, such that the topology of data is taken into account. More precisely, the median clustering has been integrated into the popular self-organizing map (SOM) [4,40] and its applicability has been demonstrated in a large scale experiment from bioinformatics [41]. Later, the same idea has been integrated into neural gas (NG) clustering together with a proof of the convergence of median SOM and median NG clustering [13]. Like K-centers clustering, the methods require an exhaustive search to obtain optimum prototypes given fixed assignments such that the complexity of a standard implementation for one epoch is $\mathcal{O}(N^2K)$ (this can be reduced to $\mathcal{O}(N^2 + NK^2)$ for median SOM, see Section 4 and [12]). Unlike K-means, local optima and overfitting can widely be avoided due to the neighborhood cooperation such that fast and reliable methods result which are robust with respect to noise in the data. Apart from this numerical stability, the methods have further benefits: they are given by simple formulas and they are very easy to implement, they rely on underlying cost functions which can be extended towards the setting of partial supervision, and in many situations a considerable speed-up of the algorithm can be obtained, as demonstrated in [12,29], for example.

In this chapter, we present an overview about neural based median clustering. We present the principle methods based on the cost functions of NG and SOM, respectively, and discuss applications, extensions and properties. Afterwards, we discuss several possibilities to speed-up the clustering algorithms, including exact methods, as well as single pass approximations for large data sets.

2   Prototype Based Clustering

Prototype based clustering aims for representing given data from some set X faithfully by means of prototypical representatives $\{\boldsymbol{w}^1, \ldots, \boldsymbol{w}^K\}$. In the standard Euclidean case, real vectors are dealt with, i.e. $X \subseteq \mathbb{R}^M$ and $\boldsymbol{w}^i \in \mathbb{R}^M$ holds for all i and some dimensionality M. For every data point $\boldsymbol{x} \in X$, the index of the *winner* is defined as the prototype

$$I(\boldsymbol{x}) = \mathrm{argmin}_j\{d(\boldsymbol{x}, \boldsymbol{w}^j)\} \tag{1}$$

where

$$d(\boldsymbol{x}, \boldsymbol{w}^j) = \sum_{i=1}^M (x_i - w_i^j)^2 \tag{2}$$

denotes the squared Euclidean distance. The *receptive field* of prototype $\boldsymbol{w}^j$ is defined as the set of data points for which it becomes winner. Typically, clustering results are evaluated by means of the *quantization error* which measures the distortion being introduced when data is represented by a prototype, i.e.

$$E := \frac{1}{2} \cdot \int \sum_{j=1}^K \delta_{I(\boldsymbol{x}),j} \cdot d(\boldsymbol{x}, \boldsymbol{w}^j) P(\boldsymbol{x}) d\boldsymbol{x} \tag{3}$$

for a given probability measure P according to which data are distributed. δ_{ij} denotes the Kronecker function. In many training settings, a finite number of data $X = \{x^1, \ldots, x^N\}$ is given in advance and the corresponding discrete error becomes

$$\hat{E} := \frac{1}{2N} \cdot \sum_{i=1}^{N} \sum_{j=1}^{K} \delta_{I(x^i),j} \cdot d(x^i, w^j) \qquad (4)$$

The popular K-means clustering algorithm aims at a direct optimization of the quantization error. In batch mode, it, in turn, determines optimum prototypes w^j given fixed assignments $I(x^i)$ and vice versa until convergence:

$$k_{ij} := \delta_{I(x^i),j}, \quad w^j := \frac{\sum_i k_{ij} x^i}{\sum_i k_{ij}} \qquad (5)$$

This update scheme is very sensitive to the initialization of prototypes such that multiple restarts are usually necessary.

Neural Gas

The self-organizing map and neural gas enrich the update scheme by neighborhood cooperation of the prototypes. This accounts for a topological ordering of the prototypes such that initialization sensitivity is (almost) avoided. The cost function of neural gas as introduced by Martinetz [46] has the form

$$E_{NG} \sim \frac{1}{2} \cdot \int \sum_{j=1}^{K} h_\sigma(\mathrm{rk}(x, w^j)) \cdot d(x, w^j) P(x) dx \qquad (6)$$

where

$$\mathrm{rk}(x, w^j) = |\{w^k \mid d(x, w^k) < d(x, w^j)\}| \qquad (7)$$

denotes the rank of prototype w^j sorted according to its distance from the data point x and $h_\sigma(t) = \exp(-t/\sigma)$ is a Gaussian shaped curve with the neighborhood range $\sigma > 0$. σ is usually annealed during training. Obviously, $\sigma \to 0$ yields the standard quantization error. For large values σ, the cost function is smoothed such that local optima are avoided at the beginning of training. For a discrete training set, this cost term becomes

$$\hat{E}_{NG} \sim \frac{1}{2N} \cdot \sum_{i=1}^{N} \sum_{j=1}^{K} h_\sigma(\mathrm{rk}(x^i, w^j)) \cdot d(x^i, w^j) \qquad (8)$$

This cost function is often optimized by means of a stochastic gradient descent. An alternative is offered by batch clustering which, in analogy to K-means, consecutively optimizes assignments and prototype locations until convergence, as described in [13]:

$$k_{ij} := \mathrm{rk}(x^i, w^j), \quad w^j := \frac{\sum_i h_\sigma(k_{ij}) x^i}{\sum_i h_\sigma(k_{ij})} \qquad (9)$$

Neighborhood cooperation takes place depending on the given data at hand by means of the ranks. This accounts for a very robust clustering scheme which is very insensitive to local optima, as discussed in [46]. Further, as shown in [47] neighborhood cooperation induces a topology on the prototypes which perfectly fits the topology of the underlying data manifold provided the sampling is sufficiently dense. Thus, browsing within this information space becomes possible.

Self-organizing Map

Unlike NG, SOM uses a priorly fixed lattice structure, often a regular low-dimensional lattice such that visualization of data can directly be achieved. Original SOM as proposed by Kohonen [39] does not possess a cost function in the continuous case, but a simple variation as proposed by Heskes does [32]. The corresponding cost function is given by

$$E_{\text{SOM}} \sim \frac{1}{2} \cdot \int \sum_{j=1}^{K} \delta_{I^*(\boldsymbol{x}),j} \cdot \sum_{k=1}^{N} h_\sigma(\text{nd}(j,l)) \cdot d(\boldsymbol{x}, \boldsymbol{w}^l) P(\boldsymbol{x}) \boldsymbol{x} \qquad (10)$$

where $\text{nd}(j,l)$ describes the distance of neurons arranged on a priorly chosen neighborhood structure of the prototypes, often a regular two-dimensional lattice, and

$$I^*(\boldsymbol{x}) = \text{argmin}_i \left\{ \sum_{l=1}^{K} h_\sigma(\text{nd}(i,l)) d(\boldsymbol{x}, \boldsymbol{w}^l) \right\} \qquad (11)$$

describes the prototype which is closest to $\boldsymbol{x}$ if averaged over the neighborhood. (This is in practice often identical to the standard winner.) In the discrete case, the cost function becomes

$$\hat{E}_{\text{SOM}} \sim \frac{1}{2N} \cdot \sum_{i=1}^{N} \sum_{j=1}^{K} \delta_{I^*(\boldsymbol{x}^i),j} \cdot \sum_{k=1}^{N} h_\sigma(\text{nd}(j,k)) \cdot d(\boldsymbol{x}^i, \boldsymbol{w}^k) \qquad (12)$$

SOM is often optimized by means of a stochastic gradient descent or, alternatively, in a fast batch mode, subsequently optimizing assignments and prototypes as follows:

$$k_{ij} := \delta_{I^*(\boldsymbol{x}^i),j}, \quad \boldsymbol{w}^j := \frac{\sum_{i,k} k_{ik} h_\sigma(\text{nd}(k,j)) \boldsymbol{x}^i}{\sum_{i,k} k_{ik} h_\sigma(\text{nd}(k,j))} \qquad (13)$$

As before, the neighborhood range σ is annealed to 0 during training, and the standard quantization error is recovered. Intermediate steps offer a smoothed cost function the optimization of which is simpler such that local optima can widely be avoided and excellent generalization can be observed. Problems can occur if the priorly chosen topology does not fit the underlying manifold and topological mismatches can be observed. Further, topological defomations can easily occur in batch optimization when annealing the neighborhood quickly as demonstrated in [16]. These problems are not present for the data optimum topology provided by NG, but, unlike SOM, NG does not offer a direct visualization of data. Batch SOM, NG, and K-means converge after a finite and usually small number of epochs [13].

3 Median Clustering

Assume data is characterized only by a matrix of pairwise nonnegative dissimilarities

$$D = (d(\boldsymbol{x}^i, \boldsymbol{x}^j))_{i,j=1,\dots,N}, \tag{14}$$

i.e. it is not necessarily contained in an Euclidan space. This setting covers several important situations in biomedical domains such as sequence data which are compared by alignment, time dependent signals for which correlation analysis gives a good dissimilarity measure, or medical images for which problem specific dissimilarity measures give good results. In particular, the data space X is discrete such that prototypes cannot be adapted smoothly within the data space.

The idea of median clustering is to restrict prototype locations to data points. The objective of clustering as stated by the quantization error (4) is well defined if prototypes are restricted to the data set $\boldsymbol{w}^j \in X$ and the squared Euclidan metric is substituted by a general term $d(\boldsymbol{x}^i, \boldsymbol{w}^j)$ given by a dissimilarity matrix only. Similarly, the cost functions of neural gas and the self-organizing map remain well-defined for $\boldsymbol{w}^j \in X$ and arbitrary terms $d(\boldsymbol{x}^i, \boldsymbol{w}^j)$. Thereby, the dissimilarities D need not fulfill the conditions of symmetry or the triangle inequality.

One can derive learning rules based on the cost functions in the same way as batch clustering by means of a subsequent optimization of prototype locations and assignments. For NG, optimization of the cost function

$$\hat{E}_{\mathrm{NG}} \sim \frac{1}{2N} \cdot \sum_{i=1}^{N} \sum_{j=1}^{K} h_\sigma(\mathrm{rk}(\boldsymbol{x}^i, \boldsymbol{w}^j)) \cdot d(\boldsymbol{x}^i, \boldsymbol{w}^j)$$

with the constraint $\boldsymbol{w}^j \in X$ yields the following algorithm for median NG:

> init
> repeat
> $\quad k_{ij} := \mathrm{rk}(\boldsymbol{x}^i, \boldsymbol{w}^j)$
> $\quad \boldsymbol{w}^j := \mathrm{argmin}_{\boldsymbol{x}^l} \sum_{i=1}^{N} h_\sigma(k_{ij}) d(\boldsymbol{x}^i, \boldsymbol{x}^l)$

Unlike batch NG, a closed solution for optimum prototype locations does not exist and exhaustive search is necessary. In consequence, one epoch has time complexity $\mathcal{O}(N^2 K)$ compared to $\mathcal{O}(NK)$ for batch NG (neglecting sorting of prototypes). Because of the discrete locations of prototypes, the probability that prototypes are assigned to the same location becomes nonzero. This effect should be avoided e.g. by means of adding random noise to the distances in every run or via an explicit collision prevention mechanism, as in [52] for the median SOM.

Similarly, median SOM can be derived from the cost function $\hat{E}_{\mathrm{SOM}}$. The following algorithm is obtained:

> init
> repeat
> $\quad k_{ij} := \delta_{I^*(\boldsymbol{x}^i),j}$
> $\quad \boldsymbol{w}^j := \mathrm{argmin}_{\boldsymbol{x}^l} \sum_{i=1}^{N} \sum_{k=1}^{K} k_{ik} h_\sigma(\mathrm{nd}(k,j)) d(\boldsymbol{x}^i, \boldsymbol{x}^l)$

As for median NG, prototypes have to be determined by exhaustive search leading to $\mathcal{O}(N^2 K)$ complexity per epoch.

Median clustering can be used for any given matrix D. It has been shown in [13] that both, median SOM and median NG converge in a finite number of steps, because both methods subsequently minimize the respective underlying cost function until they arrive at a fixed point of the algorithm. A simple demonstration of the behavior of median NG can be found in Fig. 1. The popular iris data set (see [5,15]) consists of 150 points and 3 classes.

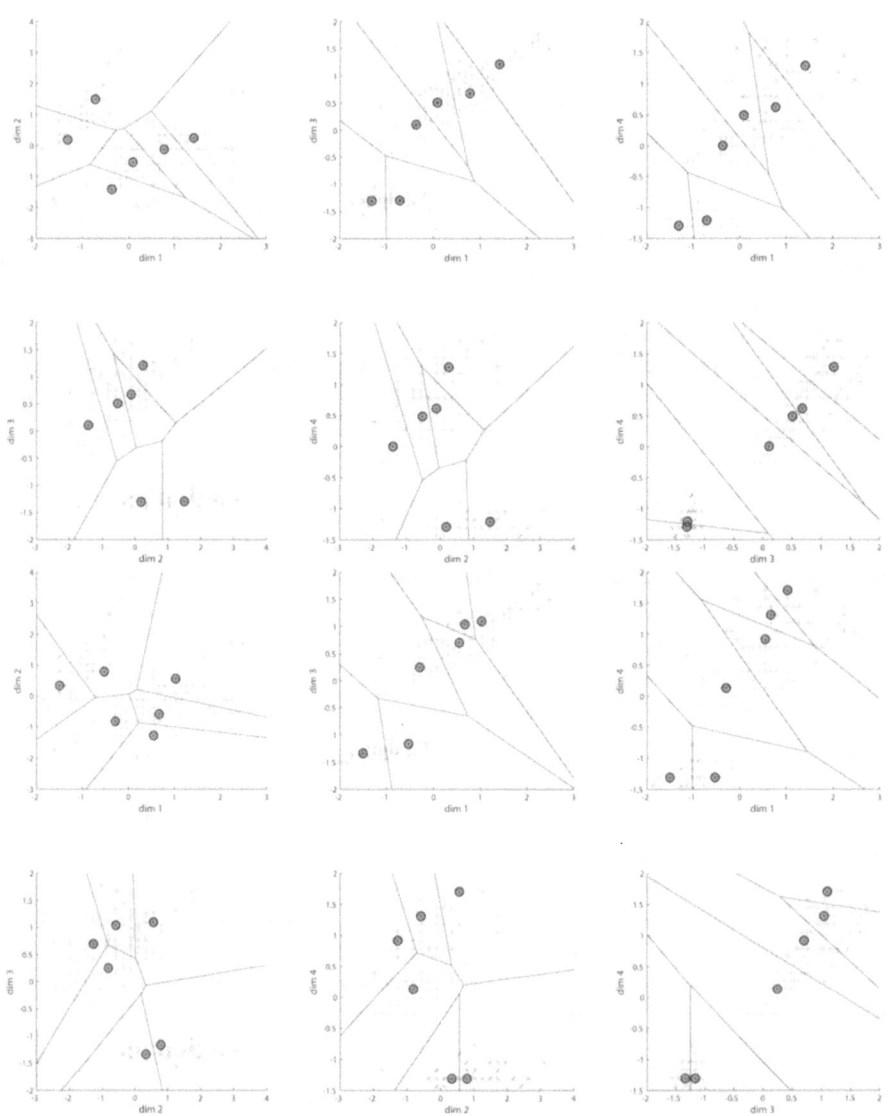

Fig. 1. Results of batch NG (top) and median NG (bottom) on the iris data set projected to two of the four data dimensions

Data are standardized to z-scores and batch NG and median NG is applied using 6 prototypes. Dimensions 3 and 4 are plotted together with the prototypes found by batch NG and median NG, respectively in Fig. 1. Obviously, the result is very similar, the quantization error of batch NG being 40.96, while median NG displays the slightly larger quantization error 44.85 due to the restriction of prototypes to data points. The classification accuracy obtained by posterior labeling is 0.84 for batch NG as compared to 0.92 for median NG. Since the prior class information is not taken into account during training, standard batch NG yields a larger classification error despite from its larger prototype flexibility.

Supervision

Often, additional label information is available for (parts of) the data set. Unsupervised data inspection and clustering does not take this information into account and cluster boundaries do not follow class distributions of the priorly known classes unless they coincide with unsupervised clusters in the data. This can be avoided by taking prior label information into account. We assume that a class label $\boldsymbol{y}^i$ is available for every data point $\boldsymbol{x}^i$. We assume $\boldsymbol{y}^i \in \mathbb{R}^d$, i.e. d classes are represented by means of a full disjunctive coding (i.e., $y_j^i = \delta_{c^i,j}$, where c^i is the index of the class for the data point $\boldsymbol{x}^i$), including the possibility of fuzzy assignments. We equip every prototype $\boldsymbol{w}^j$ with a label $\boldsymbol{Y}^j \in \mathbb{R}^d$ which is adapted during training. This vector represents the class label of the prototype, i.e. the average labels of data in its receptive field.

The aim of semisupervised clustering and data inspection is to determine prototypes and their labels in such a way that prototypes represent data point faithfully and they take the labeling into account, i.e. the prototype labels should correspond to the label of data points of its receptive field. To achieve this goal, the distance of a prototype $\boldsymbol{w}^j$ from a data point $\boldsymbol{x}^i$ is extended towards

$$d_\beta(\boldsymbol{x}^i, \boldsymbol{w}^j) := \beta \cdot d(\boldsymbol{x}^i, \boldsymbol{w}^j) + (1 - \beta) \cdot d(\boldsymbol{y}^i, \boldsymbol{Y}^j) \tag{15}$$

where $d(\boldsymbol{y}^i, \boldsymbol{Y}^j)$ denotes the squared Euclidean distance of the labels $\boldsymbol{y}^i$ and $\boldsymbol{Y}^j$, and $\beta \in (0, 1)$ balances the goal to represent the input data and the labels within the receptive field correctly.

This extended distance measure can be directly integrated into the cost functions of NG and SOM. Depending on the form of the distance measure, an extension of batch optimization or median optimization becomes possible. In both cases, the standard batch optimization scheme is accompanied by the label updates, which yields

$$\boldsymbol{Y}^j = \sum_i h_\sigma(k_{ij}) \cdot y^i / \sum_i h_\sigma(k_{ij}) \tag{16}$$

for batch and median NG, respectively, and

$$\boldsymbol{Y}^j = \sum_{ik} k_{ik} h_\sigma(\mathrm{nd}(k, j)) \cdot y^i / \sum_{ik} k_{ik} h_\sigma(\mathrm{nd}(k, j)) \tag{17}$$

for batch and median SOM, respectively. It can be shown in the same way as for standard batch and median clustering, that these supervised variants converge after a finite number of epochs towards a fixed point of the algorithm.

The effect of supervision can exemplarly be observed for the iris data set: supervised batch NG with supervision parameter $\beta = 0.5$ causes the prototypes to follow more closely the prior class borders in particular in overlapping regions and, correspondingly, an improved classification accuracy of 0.95 is obtained for supervised batch NG. The mixture parameter β constitutes a hyperparameter of training which has to be optimized according to the given data set. However, in general the sensitivity with respect to β seems to be quite low and the default value $\beta = 0.5$ is often a good choice for training. Further, supervision for only part of the training data x^i is obviously possible.

Experiments

We demonstrate the behavior of median NG for a variety of biomedical benchmark problems. NG is intended for unsupervised data inspection, i.e. it can give hints on characteristic clusters and neighborhood relationships in large data sets. For the benchmark examples, prior class information is available, such that we can evaluate the methods by means of their classification accuracy. For semisupervised learning, prototypes and corresponding class labels are directly obtained from the learning algorithm. For unsupervised training, posterior labeling of the prototypes based on a majority vote of their receptive fields can be used. For all experiments, repeated cross-validation has been used for the evaluation.

Wisconsin breast cancer. The Wisconsin breast cancer diagnostic database is a standard benchmark set from clinical proteomics [56]. It consists of 569 data points described by 30 real-valued input features: digitized images of a fine needle aspirate of breast mass are described by characteristics such as form and texture of the cell nuclei present in the image. Data are labeled by two classes, benign and malignant.

The data set is contained in the Euclidean space such that we can compare all clustering versions as introduced above for this data set using the Euclidean metric. We train 40 neurons using 200 epochs. The dataset is standardized to z-scores and randomly split into two halfs for each run. The result on the test set averaged over 100 runs is reported. We obtain a test set accuracy of 0.957 for the supervised version and 0.935 for the unsupervised version, both setting $\beta = 0.1$ which is optimum for these cases. Results for simple K-means without neighborhood cooperation yield an accuracy 0.938 for standard (unsupervised) K-means resp. 0.941 for supervised K-means. Obviously, there are only minor, though significant differences of the results of the different clustering variants on this data set: incorporation of neighborhood cooperation allows to improve K-means, incorporation of label information allows to improve fully unsupervised clustering. As expected, Euclidean clustering is superior to median versions (using the squared Euclidean norm) because the number of possible prototype locations is reduced for median clustering. However, the difference is only 1.3%, which is quite remarkable because of the comparably small data set, thus dramatically reduced flexibility of prototype locations.

The article [56] reports a test set accuracy of 0.97% using 10-fold cross-validation and a supervised learning algorithm (a large margin linear classifier including feature selection). This differs from our best classification result by 1.8%. Thereby, the goal of our approach is a faithful prototype-based representation of data, such that the result is remarkable.

Chromosomes. The Copenhagen chromosomes database is a benchmark from cytogenetics [45]. A set of 4200 human chromosomes from 22 classes (the autosomal chromosomes) are represented by the grey levels of their images and transferred to strings which represent the profile of the chromosome by the thickness of their silhouettes. This data set is non-Euclidean, consisting of strings of different length, and standard neural clustering cannot be used. Median versions, however, are directly applicable. The edit distance (also known as the Levenshtein distance [43]), is a typical distance measure for two strings of different length, as described in [35,49]. In our application, distances of two strings are computed using the standard edit distance whereby substitution costs are given by the signed difference of the entries and insertion/deletion costs are given by 4.5 [49].

The algorithms have been run using 100 neurons and 100 epochs per run. Supervised median neural gas achieves an accuracy of 0.89 for $\beta = 0.1$. This improves by 6% compared to median K-means. A larger number of prototypes allows to further improve this result: 500 neurons yield an accuracy of 0.93 for supervised median neural gas clustering and 0.91 for supervised median K-means clustering, both taken for $\beta = 0.1$. This is already close to the results of fully supervised k-nearest neighbor classification which uses all points of the training set for classification. 12-nearest neighbors with the standard edit distance yields an accuracy 0.944 as reported in [35] whereas more compact classifiers such as feedforward networks or hidden Markov models only achieve an accuracy less than 0.91, quite close to our results for only 100 prototypes.

Proteins

The evolutionary distance of 226 globin proteins is determined by alignment as described in [48]. These samples originate from different protein families: hemoglobin-α, hemoglobin-β, myoglobin, etc. Here, we distinguish five classes as proposed in [23]: HA, HB, MY, GG/GP, and others.

We use 30 neurons and 300 epochs per run. The accuracy on the test set averaged over 50 runs is reported in Fig. 2. Here, optimum mixing parameters can be observed for supervised median neural gas and $\beta \in [0.5, 0.9]$, indicating that the statistics of the inputs guides the way towards a good classification accuracy. However, an integration of the labels improves the accuracy by nearly 10% compared to fully unsupervised clustering. As beforehand, integration of neighborhood cooperation is well suited in this scenario. Unlike the results reported in [23] for SVM which uses one-versus-rest encoding, the classification in

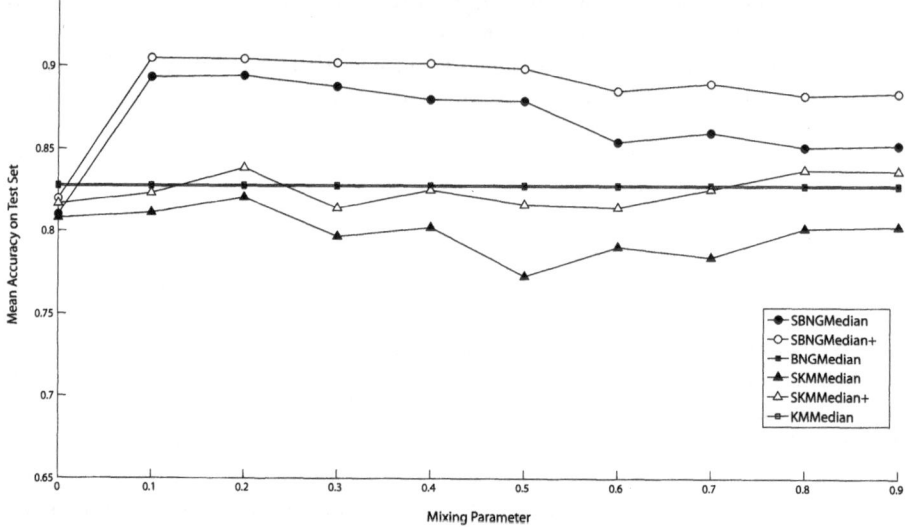

Fig. 2. Results of the methods for the protein database using alignment distance and varying mixing parameter β. The version indicated with + refers to (optimum) posterior labeling of prototypes.

our setting is given by only one clustering model. Depending on the choice of the kernel, [23] reports errors which approximately add up to 0.04 for the leave-one-out error. This result, however, is not comparable to our results due to the different error measure. A 1-nearest neighbor classifier yields an accuracy 0.91 for our setting (k-nearest neighbor for larger k is worse; [23] reports an accumulated leave-one-out error of 0.065 for 1-nearest neighbor) which is comparable to our (clustering) results.

Thus, unsupervised or semi-supervised data inspection which accounts for both, data statistics and prior labeling, reaches a classification accuracy comparable to fully supervised approaches, i.e. the clusters found by median NG are meaningful in these cases.

4   Fast Implementations

As pointed out previously, the computational cost of median clustering algorithms is quite high: the exhaustive search for the best prototypes leads to a cost of $\mathcal{O}(N^2K)$ per epoch (for both median NG and median SOM). This cost is induced by the need for evaluating a sum of the following form

$$\sum_{i=1}^{N} \alpha_{i,l,j} d(\boldsymbol{x}^i, \boldsymbol{x}^l),$$

where $\boldsymbol{x}^l$ is a candidate for prototype j. Evaluating this sum is a $\mathcal{O}(N)$ operation which has to be repeated for each candidate (N possibilities) and for each prototype (K times). As the coefficients of the sums depends on j, it might seem at first glance that there is no way to reduce the cost.

4.1 Block Summing

However, the prior structure of the SOM can be leveraged to reduce the total cost to $\mathcal{O}(N^2 + NK^2)$ per epoch [12]. Let C_j^* denote the receptive field of prototype j, more precisely

$$C_j^* := \left\{ i \in \{1, \ldots, N\} \mid I^*(\boldsymbol{x}^i) = j \right\}.$$

Then the prototype $\boldsymbol{w}^j$ is given by

$$\boldsymbol{w}^j = \arg\min_{\boldsymbol{x}^l} \sum_{k=1}^{K} h_\sigma(\mathrm{nd}(k, j)) \sum_{i \in C_k^*} d(\boldsymbol{x}^i, \boldsymbol{x}^l).$$

The main advantage of this formulation over the standard one is that there is now a clean separation between components that depend on j (the $h_\sigma(\mathrm{nd}(k, j))$ terms) and those that do not (the sums $\sum_{i \in C_k^*} d(\boldsymbol{x}^i, \boldsymbol{x}^l)$). This leads to the following version of the median SOM [12]:

> init
> repeat
> $C_j^* := \left\{ i \in \{1, \ldots, N\} \mid I^*(\boldsymbol{x}^i) = j \right\}$ (*receptive field calculation*)
> $S(k, l) := \sum_{i \in C_k^*} d(\boldsymbol{x}^i, \boldsymbol{x}^l)$ (*block summing*)
> $\boldsymbol{w}^j := \arg\min_{\boldsymbol{x}^l} \sum_{k=1}^{K} h_\sigma(\mathrm{nd}(k, j)) S(k, l)$ (*prototype calculation*)

There are $N \times K$ block sums $S(k, l)$ which can be computed in $\mathcal{O}(N^2)$ operations as the $(C_k^*)_{1 \leq k \leq N}$ form a partition of the dataset. Then the exhaustive search involves only summing K values per candidate prototype (and per prototype), leading to a total cost of $\mathcal{O}(NK^2)$ (of the same order as the computation of the receptive fields). The total computational load is therefore $\mathcal{O}(N^2 + NK^2)$. In practice, the speed up is very high. For instance, with the optimized Java implementation proposed in [12][1], a standard $\mathcal{O}(N^2 K)$ implementation of the median SOM uses approximately 5.7 seconds per epoch on the Chromosomes dataset ($N = 4200$) for $K = 100$ prototypes (arranged on a 10×10 hexagonal grid) on a standard workstation[2]. Under identical conditions, the above algorithm uses only 0.25 second per epoch while providing exactly the same results (this is 23 times faster than the standard implementation, see Table 1 for a summary of the timing results for the Median SOM variants).

Obviously, the speed up strongly depends on both N and K. For example if K is raised to 484 (for a regular hexagonal grid of size 22×22), the standard implementation of the median SOM uses approximately 30.1 seconds per epoch,

---

[1] Available at `http://gforge.inria.fr/projects/somlib/`
[2] AMD Athlon 64 3000+ processor with 2GB of main memory, running Fedora Linux 7 and with Sun 1.6 java virtual machine in server mode.

Table 1. Average time needed to complete an epoch of the Median SOM (in seconds) for the Chromosomes dataset

Algorithm	$K = 100$	$K = 484$
Standard implementation	5.74	30.1
Block Summing	0.248	3.67
Branch and bound $\Theta = \{j\}$	0.224	2.92
Branch and bound $\Theta = \{1, \ldots, K\}$	0.143	1.23
Branch and bound $\Theta = \{1, \ldots, K\}$ and early stopping	0.141	0.933

with the block summing algorithm uses only 3.67 seconds per epoch. This is still more than 8 times faster than the standard implementation, but as expected the speed up factor is worse than with a lower value of K. Nevertheless as reflected in the theoretical cost analysis, extensive simulations conducted in [12] have shown that the block summing algorithm is always faster than the standard approach which has therefore no reason to be used.

4.2   Heuristic Search

Additional reduction in the actual running time of the median SOM can be obtained via the branch and bound principle from combinatorial optimization [11]. The goal of branch and bound [42] is to avoid to perform an exhaustive search to solve a minimization problem by means of two helper procedures. The first procedure is a partition method to be applied to the search space (the *branch* part of the method). The second procedure provides *quickly* a guaranteed lower bound of the criterion to be minimized on any class of the partition of the search space (the *bound* part of the method). By *quickly* one means faster than an exhaustive evaluation of the criterion on the class.

A standard implementation of the minimization of a function f by an exhaustive search on the search space S proceeds as follows:

> initialise *best* to $s_1 \in S$
> initialise *fbest* to $f(best)$
> for all $s \in S \setminus \{s_1\}$ do
>   compute $f(s)$
>   if $f(s) < fbest$ update *best* to s and *fbest* to $f(s)$

To save some evaluations of f, a branch and bound search proceeds as follows:

> compute $C_1, \ldots, C_K$ a partition of S
> initialise *fbest* and *best* by an exhaustive search in C_1
> for $i = 2, \ldots, K$
>   compute a lower bound g for f on C_i
>   if $g < fbest$ update *best* and *fbest* by an exhaustive search in C_i

The gain comes from the possibility of pruning entire regions (classes) of the search space when the lower bound of the criterion f on such a region is higher than the best value found so far. The best gain is achieved when all regions except C_1 are pruned. Obviously, the order in which the partition classes are searched is crucial in obtaining good performances.

In the median SOM, the search space is the dataset. This provides a natural branching procedure as the receptive fields $(C_j^*)_{1 \leq j \leq K}$ of the prototypes of the SOM build a partition of the dataset. If the receptive fields have comparable sizes, i.e., around N/K, branch and bound can reduce the search space for each prototype from a size of N to a size of N/K, in the optimal case (perfect branching). This could reduce the cost of the search from $\mathcal{O}(NK^2)$ to $\mathcal{O}(NK + K^2)$. Indeed, in the ideal case, one would only evaluate the quality criterion $\sum_{k=1}^{K} h_\sigma(\mathrm{nd}(k,j))S(k,l)$ for candidate prototypes from one cluster (this would cost $\mathcal{O}(K(N/K)) = \mathcal{O}(N)$) and then compare the best value to the lower bound of each other cluster ($\mathcal{O}(K)$ additional operations).

The bounding procedure needs to provide a tight lower bound for the following quantity

$$\min_{x^l \in C_m^*} \sum_{k=1}^{K} h_\sigma(\mathrm{nd}(k,j))S(k,l).$$

A class of lower bounds is given by the following equation

$$\eta(m,j,\Theta) := \sum_{k \in \Theta} h_\sigma(\mathrm{nd}(k,j)) \min_{x^l \in C_m^*} S(k,l), \qquad (18)$$

where Θ is a subset of $\{1, \ldots, K\}$. There are several reasons for using such bounds. First the quantity $\min_{x^l \in C_m^*} S(k,l)$ depends only on k and l: it can be computed once and for all before the exhaustive search phase (in fact in parallel with the computation of the block sum $S(k,l)$ itself). The additional cost is negligible compared to other costs (there are K^2 values which are computed in $\mathcal{O}(NK)$ operations). However, computing $\eta(m,j,\{1,\ldots,K\})$ is costly, as the search process will need this bound for all m and j, leading to a total cost of $\mathcal{O}(K^3)$: this is small compared to $\mathcal{O}(N^2 + NK^2)$ but not negligible when K is large, especially compared to the best case cost ($\mathcal{O}(N^2 + NK + K^2)$ with perfect branching).

It is therefore interesting in theory to consider strict subsets of $\{1, \ldots, K\}$, in particular the singleton $\Theta = \{j\}$ which leads to the very conservative lower bound $h_\sigma(\mathrm{nd}(j,j)) \min_{x^l \in C_m^*} S(j,l)$, for which the computation cost is only $\mathcal{O}(K^2)$ for all m and j. Despite its simplicity, this bound leads to very good results in practice [11] because when the neighborhood influence is annealed during training, $h_\sigma(\mathrm{nd}(k,j))$ gets closer and closer to the Kronecker function $\delta_{k,j}$.

Compared to the improvements generated by reducing the complexity to $\mathcal{O}(N^2 + NK^2)$, the speed up provided by branch and bound is small. Under exactly the same conditions as in the previous section, the time needed per epoch is 0.22 second (compared to 0.25) when the bounds are computed with $\Theta = \{j\}$ and 0.14 second when $\Theta = \{1, \ldots, K\}$ (which shows that perfect branching does not happen as the $\mathcal{O}(K^3)$ cost of the bounds calculation does not prevent from getting a reasonable speed up). Complex additional programming tricks exposed in [11,12] can reduce even further the running time in some situation (e.g., when

K is very large), but on the Chromosomes dataset with $K = 100$, the best time is obtained with the algorithm described above. The speed up compared to a naive implementation is nevertheless quite large as the training time is divided by 40, while the results are guaranteed to be exactly identical.

When K is increased to 484 as in the previous section, the time per epoch for $\Theta = \{j\}$ is 2.92 seconds and 1.23 seconds when $\Theta = \{1, \ldots, K\}$. In this case, the "early stopping" trick described in [11,12] (see also the next Section) can be used to bring down this time to 0.93 second per epoch. This is more than 32 times faster than the naive implementation and also almost 4 times faster than the block summing method. Branch and bound together with early stopping complement therefore the block summing improvement in the following sense: when K is small (of the order of $\sqrt{N}$ or below), block summing is enough to reduce the algorithmic cost to an acceptable $\mathcal{O}(N^2)$ cost. While K grows above this limit, branch and bound and early stopping manage to reduce the influence of the $\mathcal{O}(NK^2)$ term on the total running time.

4.3   Median Neural Gas

Unfortunately the solutions described above cannot be applied to median NG, as there is no way to factor the computation of $\sum_{i=1}^{N} h_\sigma(k_{ij})d(x^i, x^l)$ in a similar way as the one used by the median SOM. Indeed, the idea is to find in the sum sub-parts that do not depend on j (the prototype) so as to re-use them for all the prototypes. The only way to achieve this goal is to use a partition on the dataset (i.e., on the index i) such that the values that depend on j, $h_\sigma(k_{ij})$, remain constant on each class. This leads to the introduction of a partition R^j whose classes are defined by

$$R_k^j = \left\{ i \in \{1, \ldots, N\} \mid \mathrm{rk}(x^i, w^j) = k \right\}.$$

This is in fact the partition induced by the equivalence relation on $\{1, \ldots, N\}$ defined by $i \sim_j i'$ if and only if $h_\sigma(k_{ij}) = h_\sigma(k_{i'j})$. Using this partition, we have

$$\sum_{i=1}^{N} h_\sigma(k_{ij})d(x^i, x^l) = \sum_{k=1}^{K} h_\sigma(k) \sum_{i \in R_k^j} d(x^i, x^l). \tag{19}$$

At first glance, this might look identical to the factorisation used for the median SOM. However there is a crucial difference: here the partition **depends** on j and therefore the block sums $\sum_{i \in R_k^j} d(x^i, x^l)$ cannot be precomputed (this factorization will nevertheless prove very useful for early stopping).

For both algorithms, the fast decrease of h_σ suggests an approximation in which small values of $\alpha_{i,l,j}$ (i.e., of $h_\sigma(k_{ij})$ and of $h_\sigma(\mathrm{nd}(k, j))$) are discarded. After a few epochs, this saves a lot of calculation but the cost of initial epochs remains unchanged. Moreover, this approximation scheme changes the results of the algorithms, whereas the present section focuses on exact and yet fast implementation of the median methods.

A possible source of optimization for median NG lies in the so called "early stopping" strategy exposed for the median SOM in [11,12]. The idea is to leverage the fact that the criterion to minimize is obtained by summing positive values. If the sum is arranged in such as way that large values are added first, the partial result (i.e., the sum of the first terms of the criterion) can exceed the best value obtained so far (from another candidate). Then the loop that implements the summing can be stopped prematurely reducing the cost of the evaluation of the criterion. Intuitively, this technique works if the calculation are correctly ordered: individual values in the sum should be processed in decreasing order while candidate prototypes should be tested in order of decreasing quality.

For the median SOM, as shown in [11,12], early stopping, while interesting in theory, provides speed up only when K is large as the use of block summing has already reduced the cost of the criterion evaluation from $\mathcal{O}(N)$ to $\mathcal{O}(K)$. On the Chromosomes dataset for instance, the previous Section showed that early stopping gains nothing for $K = 100$ and saves approximately 24% of the running time for $K = 484$ (see Table 1).

However, as there is no simplification in the criterion for median NG, early stopping could cause some improvement, especially as the sorting needed to compute the rk function suggests an evaluation order for the criterion and an exploration order during the exhaustive search.

The computation of $w^j := \arg\min_{x^l} \sum_{i=1}^{N} h_\sigma(k_{ij})d(x^i, x^l)$ by an early stopping algorithm takes the following generic form:

$q := \infty$
for $l \in \{1, \ldots, N\}$ *(candidate loop)*
    $s := 0$
    for $i \in \{1, \ldots, N\}$ *(inner loop)*
        $s := s + h_\sigma(k_{ij})d(x^i, x^l)$
        if $s > q$ break the inner loop
    endfor *(inner loop)*
    if $s < q$ then
        $q := s$
        $w^j := x^l$
    endif
endfor *(candidate loop)*

Ordering. Both loops (*candidate* and *inner*) can be performed in specific orders. The *candidate loop* should analyse the data points x^l in decreasing quality order, i.e., it should start with x^l such that $\sum_{i=1}^{N} h_\sigma(k_{ij})d(x^i, x^l)$ is small and end with the points that have a large value of this criterion. This optimal order is obviously out of reach because computing it implies the evaluation of all the values of the criterion, precisely what we want to avoid. However, Median Neural Gas produces clusters of similar objects, therefore the best candidates for prototype w^j are likely to belong to the receptive field of this prototype at the previous epoch. A natural ordering consists therefore in trying first the elements of this receptive field and then elements of the receptive fields of other prototypes in increasing order of dissimilarities between prototypes. More precisely, with

$$C_k := \left\{ i \in \{1, \ldots, N\} \mid I(\boldsymbol{x}^i) = k \right\},$$

the *candidate loop* starts with $C_{j_1} = C_j$ and proceeds through the $(C_{j_i})_{2 \le i \le K}$ with $d(\boldsymbol{w}^j, \boldsymbol{w}^{j_{i-1}}) \le d(\boldsymbol{w}^j, \boldsymbol{w}^{j_i})$. Computing this order is fast ($\mathcal{O}(K^2 \log K)$ operations for the complete prototype calculation step).

Another solution consists in ordering the *candidate loop* according to increasing ranks k_{lj} using the partition R^j defined previously instead of the receptive fields.

The *inner loop* should be ordered such that s increases as quickly as possible: high values of $h_\sigma(k_{ij})d(\boldsymbol{x}^i, \boldsymbol{x}^l)$ should be added first. As for the *candidate loop* this cannot be achieved exactly without loosing all benefits of the early stopping. A simple solution consists again in using the partition R^j. The rational is that the value of $h_\sigma(k_{ij})d(\boldsymbol{x}^i, \boldsymbol{x}^l)$ is likely to be dominated by the topological term $h_\sigma(k_{ij})$, especially after a few epochs when h_σ becomes peaked in 0. This ordering has three additional benefits. As explained above, the factorized representation provided by equation (19) allows to save some calculations. Moreover, R^j can be computed very efficiently, as a side effect of computing the ranks k_{ij}. If they are stored in a NK array, R^j is obtained by a single pass on the index set $\{1, \ldots, N\}$. Computing the R^j for all j has therefore a $\mathcal{O}(NK)$ cost. Finally, R^j can also be used for ordering the *candidate loop*.

It should be noted that using different orderings for the *candidate loop* can lead to different final results for the algorithm in case of ties between candidates for prototypes. In practice, the influence of such differences is extremely small, but contrarily to the SOM for which all experiments produced in [11,12] and in this chapter gave exactly the same results, regardless of the actual implementation, variants of the Median Neural Gas exhibit a small variability in the results (less than one percent of differences in the quantization error, for instance). Those differences have been neglected as they could be suppressed via a slightly more complex implementation in which ties between candidate prototypes are broken at random (rather than via the ordering); using the same random generator would produce exactly the same results in all implementations.

Early stopping granularity. Experiments conducted in [11,12] have shown that early stopping introduces a non negligible overhead to the *inner loop* simply because it is the most intensive part of the algorithm which is executed N^2K times in the worst case. A coarse grain early stopping strategy can be used to reduce the overhead at the price of a more complex code and of less early stops. The idea is to replace the standard *inner loop* by the following version:

$$s := 0$$
for $m \in \{1, \ldots, M\}$ *(monitoring loop)*
    for $i \in B_m$ *(internal loop)*
        $s := s + h_\sigma(k_{ij})d(\boldsymbol{x}^i, \boldsymbol{x}^l)$
    endfor *(internal loop)*
    if $s > q$ break the monitoring loop
endfor *(monitoring loop)*

Table 2. Average time needed to complete an epoch of the Median Neural Gas (in seconds) for the Chromosomes dataset

Algorithm	$K = 100$	$K = 500$
Standard implementation	5.31	25.9
Early stopping without ordering	5.26	24.3
Early stopping with *candidate loop* ordering	4.38	20.7
Full ordering and fine grain early stopping	1.16	7.45
Full ordering and coarse grain early stopping	0.966	5.91

The main device is a partition $B = (B_1, \ldots, B_M)$ of $\{1, \ldots, N\}$ which is used to divide the computation into uninterrupted calculations (*internal loops*) and to check on a periodic basis by the *monitoring loop*. The value of M can be used to tune the grain of the early stopping with a classical trade-off between the granularity of the monitoring and its overhead. In this chapter, we have focused on a particular way of implementing this idea: rather than using an arbitrary partition B, we used the R^j partition. It has two advantages over an arbitrary one: is provides an interesting ordering of the *monitoring loop* (i.e., in decreasing order of $h_\sigma(k_{ij})$) and allows the code to use the factorized equation (19).

Experiments. Variants of the early stopping principle applied to Median Neural Gas were tested on the Chromosomes ($N = 4200$) with $K = 100$ and $K = 500$. They are summarized in Table 2. The standard implementation corresponds to the exhaustive search $\mathcal{O}(N^2 K)$ algorithm. The need for ordering is demonstrated by the results obtained by a basic implementation of early stopping in which the natural data ordering is used for both *candidate* and *inner* loops. Moreover, while the *candidate loop* order based on receptive fields reduces the running time, the gain remains limited when the *inner* loop is not ordered (the running time is reduced by approximately 20% compared to the standard implementation).

Much better improvements are reached when the R^j partitions are used to order both loops. The running time is divided by more than 5 for $K = 100$ and by more than 4 for $K = 500$, when a coarse grain early stopping method is used. The fine grain version is slightly less efficient because of the increased overhead in the *inner loop*.

The structure of the Median Neural Gas algorithm prevents the use of the block summing trick which is the main source of improvement for the Median SOM. In the case of Neural Gas, early stopping provides better improvement over the state-of-the-art implementation, than it does for the SOM, because it targets an internal loop with $\mathcal{O}(N)$ complexity whereas the block summing approach leads to a $\mathcal{O}(K)$ inner loop. In the end, the optimized Median SOM remains much faster than the Median Neural Gas (by a factor 6). However, the SOM is also very sensitive to its initial configuration whereas Neural Gas is rather immune to this problem. In practice, it is quite common to restart the SOM several times from different initial configuration, leading to quite comparable running time for both methods.

5   Approximate Patch Clustering for Large Data Sets

A common challenge today [57], arising especially in computational biology, image processing, and physics, are huge datasets whose pairwise dissimilarities cannot be hold at once within random-access memory during computation, due to the sheer amount of data (a standard workstation with 4 GB of main memory cannot hold more than $N = 2^{15}$ data points when they are described by a symmetric dissimilarity matrix). Thus, data access is costly and only a few, ideally at most one pass through the data set is still affordable.

Most work in this area can be found in the context of heuristic (possibly hierarchical) clustering on the one side and classical K-means clustering on the other side. Heuristic algorithms often directly assign data consecutively to clusters based on the distance within the cluster and allocate new clusters as required. Several popular methods include CURE, STING, and BIRCH [22,55,58]. These methods do not rely on a cost function such that an incorporation of label information into the clustering becomes difficult.

Extensions of K-means clustering can be distinguished into methods which provide guarantees on the maximum difference of the result from classical K-means, such as presented in the approaches [21,34]. However, these variants use resources which scale in the worst case with a factor depending on N (N being the number of points) with respect to memory requirements or passes through the data set. Alternatives are offered by variants of K-means which do not provide approximation guarantees, but which can be strictly limited with respect to space requirements and time. An early approach has been proposed in [50]: data are clustered consecutively in small patches, whereby the characteristics of the data and the possibility to compress subsets of data are taken into account. A simpler although almost as efficient method has been proposed in [14]: Standard K-means is performed consecutively for patches of the data whereby each new patch is enriched by the prototypes obtained in the previous patch. A sufficient statistics of the outcome of the last run can thereby easily be updated in a consecutive way, such that the algorithm provides cluster centres after only one pass through the data set, thereby processing the data consecutively in patches of predefined fixed size.

Some of these ideas have been transferred to topographic maps: the original median SOM [41] proposes simple sampling to achieve efficient results for huge data sets. Simple sampling is not guaranteed to preserve the statistics of the data and some data points might not be used for training at all, because of which reason [41] proposes to use all data in the last run. An approach which uses all available statistics consists in an extension of patch clustering towards neural gas and alternatives, as proposed in [2,3]. This method processes data in patches, thereby integrating the sufficient statistics of results of the previous run, such that all available information processed so far is used in each consecutive clustering step. Since the runs rely on a statistics of the data, the overall result only approximates the optimum solution obtained by standard batch clustering. However, in practice, results are quite good.

An extension of this method to median clustering is possible and yields promising results, as proposed in [29]. Here we give an introduction to this simple and powerful extension of median clustering to huge data sets. As a side effect, this method dramatically reduces the complexity of clustering to linear complexity in N, N being the number of data points.

Assume, as before, a dissimilarity matrix D with entries d_{ij} representing the dissimilarity of patterns. Here we assume symmetry of the matrix, but no further requirements need to be fulfilled. For huge data sets, only parts of the matrix D fit into main memory. We assume that access to single elements of the matrix is possible at any time, e.g. the elements are stored in a database or the dissimilarity values are computed on the fly by means of some (possibly complex) dissimilarity measure (such as pairwise alignment of proteins using FASTA). The exact way how dissimilarities are accessed is not relevant for patch clustering.

During processing of patch Median NG, n_p disjoint patches of fixed size $p = \lfloor m/n_p \rfloor$ are taken from the dissimilarity matrix D consecutively,[3] where every patch

$$P_i = (d_{st})_{s,t=(i-1)\cdot p,\ldots,i\cdot p-1} \in \mathbb{R}^{p\times p}$$

is a submatrix of D, representing data points $(i-1)\cdot p$ to $i\cdot p-1$. The patches are small such that they fit into main memory. The idea of the patch scheme is to add the prototypes from the processing of the former patch P_{i-1} as additional datapoints to the current patch P_i, forming an extended patch P_i^* to work on further. The additional datapoints – the former prototypes – are weighted according to the size of their receptive fields, i.e. how many datapoints they have been representing in the former patch. Therefore, every datapoint x^i, as a potential prototype, is equipped with a multiplicity m_i, that is at first initialized with $m_i = 1$. Unlike simple sampling strategies, every point of the dataset is considered exactly once and a sufficient statistics of all already processed data is passed to further patches by means of the weighted prototypes.

Unlike the situation of patch NG in Euclidean space [3,2], where inter-patch distances can always be recalculated with help of the Euclidean metric, we are now dealing with an unknown mathematical space. We have to construct the extended patch from given dissimilarity data. The extended patch P_i^* is defined as

$$P_i^* = \begin{pmatrix} d(N_{i-1}) & d(N_{i-1}, P_i) \\ \hline d(N_{i-1}, P_i)^T & P_i \end{pmatrix}$$

---

[3] The remainder is no further considered here for simplicity. In the practical implementation the remaining datapoints are simply distributed over the first $(M - p\cdot n_p)$ patches.

where

$$d(N_{i-1}) = (d_{st})_{s,t \in N_{i-1}} \in \mathbb{R}^{K \times K}$$

$$d(N_{i-1}, P_i) = (d_{st})_{s \in N_{i-1}, t=(i-1) \cdot p, \ldots, i \cdot p-1} \in \mathbb{R}^{K \times p}$$

denote the inter-distances of former prototypes and the distances between former prototypes and current patch points, respectively. Every point is weighted with a multiplicity m_j which is set to 1 for all new points $j \in [(i-1) \cdot p, i \cdot p - 1]$. For points which stem from prototypes, the multiplicity is set to the sum of the multiplicities of all points in its receptive field.

To apply median clustering, we have to incorporate these multiplicities into the learning scheme. The cost function becomes

$$\hat{E}_{NG} \sim \frac{1}{2N} \cdot \sum_{i=1}^{N} \sum_{j=1}^{K} h_\sigma(\mathrm{rk}(\boldsymbol{x}^i, \boldsymbol{w}^j)) \cdot m_j \cdot d(\boldsymbol{x}^i, \boldsymbol{w}^j) \qquad (20)$$

where, as before, prototype locations $\boldsymbol{w}^j$ are restricted to data points. Optimum ranks are obtained as beforehand. Optimum prototypes are determined by means of the formula

$$\boldsymbol{w}^j = \mathrm{argmin}_{\boldsymbol{x}^l} \sum_{i=1}^{N} h_\sigma(r_{ij}) \cdot m_j \cdot d(\boldsymbol{x}^i, \boldsymbol{x}^l) \qquad (21)$$

Picking up the pieces, we obtain the following algorithm:

Patch Median Neural Gas

Cut the first Patch P_1
Apply Median NG on $P_1 \longrightarrow$ Prototypes N_1
Update Multiplicities m_j

Repeat for $i = 2, \ldots, n_p$
    Cut patch P_i
    Construct Extended Patch P_i^* using P_i and N_{i-1}
    Apply modified Median NG with Multiplicities
        $\longrightarrow$ Prototypes N_i
    Update Multiplicities m_j

Return final Prototypes N_{n_p}

Median SOM can be extended to patch clustering in a similar way.

We demonstrate the behavior of patch clustering on the breast cancer data set which has been used beforehand. Here, we compare data points with the Cosine Measure

$$d_{cos}(\boldsymbol{x}^i, \boldsymbol{x}^j) = 1 - \frac{\boldsymbol{x}^i \cdot \boldsymbol{x}^j}{\|\boldsymbol{x}^i\|_2 \cdot \|\boldsymbol{x}^j\|_2}.$$

Standard median batch NG for 40 neurons and 100 epochs yields an average classification accuracy of 0.95 in a repeated 10-fold cross-validation. In comparison, patch median NG with 5 patches, i.e. 114 data points per patch, arrives at a classification accuracy of 0.94, yielding only a slight decrease of the classification accuracy.

The complexity of standard median NG is improved by incorporation of patches, as can be seen as follows. Assume a fixed patch size p independent of the number of datapoints, e.g. p is chosen according to the main memory. Then the algorithm uses only $\mathcal{O}(\frac{m}{p} \cdot (p + K)^2) = \mathcal{O}(m \cdot p + m \cdot K) = \mathcal{O}(m)$ entries of the dissimilarity matrix, compared to $\mathcal{O}(m^2)$ in the original Median NG method. Moreover, every epoch (within a patch) has complexity $\mathcal{O}(p^2) = $ constant as opposed to $\mathcal{O}(N^2)$ for an epoch in full median clustering. Therefore the method does not only overcome the problem of limited memory, it also dramatically accelerates the processing of datasets, what might be useful in time critical applications.

6   Discussion

Neural clustering methods such as SOM and NG offer robust and flexible tools for data inspection. In biomedical domains, data are often nonvectorial such that extensions of the original methods towards general dissimilarity data have to be used. In this chapter, we presented an overview about one particularly interesting technique which extends NG and SOM towards dissimilarities by means of the generalized median. Prototypes are restricted to data positions such that the standard cost functions are well defined and extensions such as supervision can easily be transferred to this setting. Moreover, this way, clusters are represented in terms of typical exemplars from the data set, i.e. the idea offers a data representation which can be easily interpreted by experts in biomedical domains.

These benefits are paid back by increased costs in a naive implementation of the algorithms, the complexity of one epoch being of order N^2 instead of N, where N refers to the number of data points. Since data are represented by a general $N \times N$ dissimilarity matrix instead of N single vectors, these increased costs are to some extent unavoidable if the full information contained in the data is considered. Nevertheless, a variety of structural aspects allow to reduce the costs of median clustering in practical situations.

We discussed a variety of techniques which lead to a dramatic decrease of the training time while (approximately) preserving the quality of the original methods. These approaches can be decomposed into exact methods which provably lead to the same results as the original implementation and approximations which slightly reduce the quality of the results in return for an improved efficiency. Exact methods include

- block summing for median SOM due to the specific and fixed structure of the SOM neighborhood; as pointed out in this chapter, block summing leads to a major reduction of the computation time in this case. The method cannot be applied to NG, though, because NG does not rely on a priorly fixed lattice structure.
- branch and bound methods which allow to reduce the number of necessary computations depending on the situation at hand; usually, the computational

savings strongly depend on the order in which computations are performed. As pointed out in this chapter, branching can be done with respect to candidate prototypes on the one hand and summands which contribute to the overall cost associated with one prototype on the other hand. For both settings, the topological ordering of the data suggests a natural decomposition of the whole search space into parts. This procedure yields to a significant reduction of the computational costs for NG in particular for later states of training with partially ordered setting. For SOM, the savings are only minor compared to savings by means of block summing, though possibly significant depending on the number of prototypes.

These methods lead to the same results as a naive implementation but run in a fraction of the time.

Compared to these approaches, approximate methods constitute a compromise of accuracy and complexity. We presented a patch clustering approach for median clustering, which processes data in patches of fixed size and integrates the results by means of the sufficient statistics of earlier runs. This way, the computation time is reduced from $\mathcal{O}(N^2)$ to $\mathcal{O}(N)$. In particular, only a small part of the dissimilarity matrix is considered in patch training. This has the additional benefit that, this way, only a finite and fixed memory size is required and the clustering method can readily be applied to huge streaming data sets. Further, since only a fraction of the dissimilarity matrix needs to be computed, this method is particularly suited for biomedical applications with complex dissimilarity measures such as alignment distance.

References

1. Al-Harbi, S., Rayward-Smith, V.: The use of a supervised k-means algorithm on real-valued data with applications in health. In: Chung, P.W.H., Hinde, C.J., Ali, M. (eds.) IEA/AIE 2003. LNCS, vol. 2718, pp. 575–581. Springer, Heidelberg (2003)
2. Alex, N., Hammer, B.: Parallelizing single pass patch clustering. In: Verleysen, M. (ed.) ESANN 2008, pp. 227–232 (2008)
3. Alex, N., Hammer, B., Klawonn, F.: Single pass clustering for large data sets. In: Proceedings of 6th International Workshop on Self-Organizing Maps (WSOM 2007), Bielefeld, Germany, September 3-6 (2007)
4. Ambroise, C., Govaert, G.: Analyzing dissimilarity matrices via Kohonen maps. In: Proceedings of 5th Conference of the International Federation of Classification Societies (IFCS 1996), Kobe (Japan), March 1996, vol. 2, pp. 96–99 (1996)
5. Anderson, E.: The irises of the gaspe peninsula. Bulletin of the American Iris Society 59, 25 (1935)
6. Arora, S., Raghavan, P., Rao, S.: Approximation schemes for euclidean k-medians and related problems. In: Proceedings of the 30th Annual ACM Symposium on Theory of Computing, pp. 106–113 (1998)
7. Barreto, G.A.: Time series prediction with the self-organizing map: A review. In: Hammer, B., Hitzler, P. (eds.) Perspectives on Neural-Symbolic Integration. Springer, Heidelberg (2007)

8. Boulet, R., Jouve, B., Rossi, F., Villa, N.: Batch kernel som and related laplacian methods for social network analysis. In: Neurocomputing (2008) (to be published)
9. Celeux, G., Diday, E., Govaert, G., Lechevallier, Y., Ralambondrainy, H.: Classification Automatique des Données. Bordas, Paris (1989)
10. Charikar, M., Guha, S., Tardos, A., Shmoys, D.B.: A constant-factor approcimation algorithm for the k-median problem. Journal of Computer and System Sciences 65, 129 (2002)
11. Conan-Guez, B., Rossi, F.: Speeding up the dissimilarity self-organizing maps by branch and bound. In: Sandoval, F., Prieto, A.G., Cabestany, J., Graña, M. (eds.) IWANN 2007. LNCS, vol. 4507, pp. 203–210. Springer, Heidelberg (2007)
12. Conan-Guez, B., Rossi, F., El Golli, A.: Fast algorithm and implementation of dissimilarity self-organizing maps. Neural Networks 19(6-7), 855–863 (2006)
13. Cottrell, M., Hammer, B., Hasenfuss, A., Villmann, T.: Batch and median neural gas. Neural Networks 19, 762–771 (2006)
14. Farnstrom, F., Lewis, J., Elkan, C.: Scalability for clustering algorithms revisited. SIGKDD Explorations 2(1), 51–57 (2000)
15. Fisher, R.A.: The use of multiple measurements in axonomic problems. Annals of Eugenics 7, 179–188 (1936)
16. Fort, J.-C., Letrémy, P., Cottrell, M.: Advantages and drawbacks of the batch kohonen algorithm. In: Verleysen, M. (ed.) ESANN 2002, pp. 223–230. D Facto (2002)
17. Frey, B., Dueck, D.: Clustering by passing messages between data points. Science 315, 972–977 (2007)
18. Frey, B., Dueck, D.: Response to clustering by passing messages between data points. Science 319, 726d (2008)
19. Graepel, T., Herbrich, R., Bollmann-Sdorra, P., Obermayer, K.: Classification on pairwise proximity data. In: NIPS, vol. 11, pp. 438–444. MIT Press, Cambridge (1999)
20. Graepel, T., Obermayer, K.: A stochastic self-organizing map for proximity data. Neural Computation 11, 139–155 (1999)
21. Guha, S., Mishra, N., Motwani, R., O'Callaghan, L.: Clustering data streams. In: IEEE Symposium on Foundations of Computer Science, pp. 359–366 (2000)
22. Guha, S., Rastogi, R., Shim, K.: Cure: an efficient clustering algorithm for large datasets. In: Proceedings of ACM SIGMOD International Conference on Management of Data, pp. 73–84 (1998)
23. Haasdonk, B., Bahlmann, C.: Learning with distance substitution kernels. In: Pattern Recognition - Proc. of the 26th DAGM Symposium (2004)
24. Hammer, B., Hasenfuss, A.: Relational neural gas. In: Hertzberg, J., Beetz, M., Englert, R. (eds.) KI 2007. LNCS, vol. 4667, pp. 190–204. Springer, Heidelberg (2007)
25. Hammer, B., Jain, B.J.: Neural methods for non-standard data. In: Verleysen, M. (ed.) European Symposium on Artificial Neural Networks 2004, pp. 281–292. D-side publications (2004)
26. Hammer, B., Micheli, A., Sperduti, A., Strickert, M.: Recursive self-organizing network models. Neural Networks 17(8-9), 1061–1086 (2004)
27. Hammer, B., Villmann, T.: Classification using non standard metrics. In: Verleysen, M. (ed.) ESANN 2005, pp. 303–316. d-side publishing (2005)
28. Hansen, P., Mladenovic, M.: Todo. Location Science 5, 207 (1997)
29. Hasenfuss, A., Hammer, B.: Single pass clustering and classification of large dissimilarity datasets. In: AIPR (2008)

30. Hathaway, R.J., Bezdek, J.C.: Nerf c-means: Non-euclidean relational fuzzy clustering. Pattern Recognition 27(3), 429–437 (1994)
31. Hathaway, R.J., Davenport, J.W., Bezdek, J.C.: Relational duals of the c-means algorithms. Pattern Recognition 22, 205–212 (1989)
32. Heskes, T.: Self-organizing maps, vector quantization, and mixture modeling. IEEE Transactions on Neural Networks 12, 1299–1305 (2001)
33. Hofmann, T., Buhmann, J.M.: Pairwise data clustering by deterministic annealing. IEEE Transactions on Pattern Analysis and Machine Intelligence 19(1), 1–14 (1997)
34. Jin, R., Goswami, A., Agrawal, G.: Fast and exact out-of-core and distributed k-means clustering. Knowledge and Information System 1, 17–40 (2006)
35. Juan, A., Vidal, E.: On the use of normalized edit distances and an efficient k-nn search technique (k-aesa) for fast and accurate string classification. In: ICPR 2000, vol. 2, pp. 680–683 (2000)
36. Kaski, S., Nikkilä, J., Oja, M., Venna, J., Törönen, P., Castren, E.: Trustworthiness and metrics in visualizing similarity of gene expression. BMC Bioinformatics 4 (2003)
37. Kaski, S., Nikkilä, J., Savia, E., Roos, C.: Discriminative clustering of yeast stress response. In: Seiffert, U., Jain, L., Schweizer, P. (eds.) Bioinformatics using Computational Intelligence Paradigms, pp. 75–92. Springer, Heidelberg (2005)
38. Kaufman, L., Rousseeuw, P.J.: Clustering by means of medoids. In: Dodge, Y. (ed.) Statistical Data Analysis Based on the L1-Norm and Related Methods, pp. 405–416. North-Holland, Amsterdam (1987)
39. Kohonen, T.: Self-Organizing Maps. Springer, Heidelberg (1995)
40. Kohonen, T.: Self-organizing maps of symbol strings. Technical report A42, Laboratory of computer and information science, Helsinki University of technology, Finland (1996)
41. Kohonen, T., Somervuo, P.: How to make large self-organizing maps for nonvectorial data. Neural Networks 15, 945–952 (2002)
42. Land, A.H., Doig, A.G.: An automatic method for solving discrete programming problems. Econometrica 28, 497–520 (1960)
43. Levenshtein, V.I.: Binary codes capable of correcting deletions, insertions and reversals. Sov. Phys. Dokl. 6, 707–710 (1966)
44. Lu, Y., Lu, S., Fotouhi, F., Deng, Y., Brown, S.: Incremental genetic k-means algorithm and its application in gene expression data analysis. BMC Bioinformatics 5, 172 (2004)
45. Lundsteen, C., Phillip, J., Granum, E.: Quantitative analysis of 6985 digitized trypsin G-banded human metaphase chromosomes. Clinical Genetics 18, 355–370 (1980)
46. Martinetz, T., Berkovich, S., Schulten, K.: 'neural-gas' network for vector quantization and its application to time-series prediction. IEEE Transactions on Neural Networks 4, 558–569 (1993)
47. Martinetz, T., Schulten, K.: Topology representing networks. Neural Networks 7(507-522) (1994)
48. Mevissen, H., Vingron, M.: Quantifying the local reliability of a sequence alignment. Protein Engineering 9, 127–132 (1996)
49. Neuhaus, M., Bunke, H.: Edit distance-based kernel functions for structural pattern classification. Pattern Recognition 39(10), 1852–1863 (2006)
50. Bradley, P.S., Fayyad, U., Reina, C.: Scaling clustering algorithms to large data sets. In: Proceedings of the Fourth International Conference on Knowledge Discovery and Data Mining, pp. 9–15. AAAI Press, Menlo Park (1998)

51. Qin, A.K., Suganthan, P.N.: Kernel neural gas algorithms with application to cluster analysis. In: ICPR 2004, vol. 4, pp. 617–620 (2004)
52. Rossi, F.: Model collisions in the dissimilarity SOM. In: Proceedings of XVth European Symposium on Artificial Neural Networks (ESANN 2007), Bruges (Belgium), pp. 25–30 (April 2007)
53. Shamir, R., Sharan, R.: Approaches to clustering gene expression data. In: Jiang, T., Smith, T., Xu, Y., Zhang, M.Q. (eds.) Current Topics in Computational Biology. MIT Press, Cambridge (2001)
54. Villmann, T., Seiffert, U., Schleif, F.-M., Brüß, C., Geweniger, T., Hammer, B.: Fuzzy labeled self-organizing map with label-adjusted prototypes. In: Schwenker, F., Marinai, S. (eds.) ANNPR 2006. LNCS, vol. 4087, pp. 46–56. Springer, Heidelberg (2006)
55. Wang, W., Yang, J., Muntz, R.: Sting: a statistical information grid approach to spatial data mining. In: Proceedings of the 23rd VLDB Conference, pp. 186–195 (1997)
56. Wolberg, W., Street, W., Heisey, D., Mangasarian, O.: Computer-derived nuclear features distinguish malignant from benign breast cytology. Human Pathology 26, 792–796 (1995)
57. Yang, Q., Wu, X.: 10 challenging problems in data mining research. International Journal of Information Technology & Decision Making 5(4), 597–604 (2006)
58. Zhang, T., Ramakrishnan, R., Livny, M.: Birch: an efficient data clustering method for very large databases. In: Proceedings of the 15th ACM SIGACT-SIGMOD-SIGART Symposium on Principles of Databas Systems, pp. 103–114 (1996)

Visualization of Structured Data via Generative Probabilistic Modeling

Nikolaos Gianniotis and Peter Tiňo

School of Computer Science, University of Birmingham,
Birmingham B15 2TT, United Kingdom

Abstract. We propose a generative probabilistic approach to constructing topographic maps of sequences and tree-structured data. The model formulation specifies a low-dimensional manifold of local noise models on the structured data. The manifold of noise models is induced by a smooth mapping from a low dimensional Euclidean latent space to the parameter space of local noise models. In this paper, we consider noise models endowed with hidden Markovian state space structure, namely Hidden Markov Tree Models (HMTM) and Hidden Markov Models (HMM). Compared with recursive extensions of the traditional Self-Organizing Map that can be used to visualize sequential or tree-structured data, topographic maps formulated within this framework possess a number of advantages such as a well defined cost function that drives the model optimization, the ability to test for overfitting and the accommodation of alternative local noise models implicitly expressing different notions of structured data similarity. Additionally, using information geometry one can calculate magnification factors on the constructed topographic maps. Magnification factors are a useful tool for cluster detection in non-linear topographic map formulations. We demonstrate the framework on two artificial data sets and chorals by J.S. Bach represented as sequences, as well as on images represented as trees.

1 Introduction

Topographic mapping [1,2,3,4] is the data processing technique of constructing maps that capture relationships between the data items in a given dataset. In geographical maps the affinities between objects on the map are in correspondence to the distances between the real world objects represented on the map. In topographic maps of data, distances between data reflect the similarity/closeness of the data items which may be Euclidean, Mahalanobis, statistical etc. Clearly topographic mapping is a data visualization technique when the constructed map is a two- or three-dimensional map, but the term data visualization[1] is much broader. For example in [5] an approach to visualizing data structures is presented that is based on the adeptness of the human visual system of observing large numbers of branches and leaves on a botanical tree. The approach is

---

[1] Nevertheless, after this clarification we shall interchangeably use the term "visualization" instead of the more accurate "topographic mapping".

M. Biehl et al.: (Eds.): Similarity-Based Clustering, LNAI 5400, pp. 118–137, 2009.

demonstrated on a file system by adopting a botanical representation where files, directories etc. are represented by elements such as branches or leaves.

In this work we are not concerned with representational issues such as the adoption of suitable color schemes, icons or graphics that consider particular aspects of the human visual system when constructing topographic maps. In our construction of a map, data items are simply represented as points. The crux of the work is to endow topographic maps with a clear understanding of *why* data points are mapped to their particular locations and *how* to interpret the distances between them. The data items that will concern us here have structured types.

The Self Organizing Map (SOM) [1] is one of the archetype algorithms used in topographic mapping and is a paradigm that has inspired numerous extensions. One avenue pursued by extensions of SOM is the processing of structured data. Some approaches provide the neurons with feed-back connections that allow for natural processing of recursive data types. Important models that belong to this class of extensions are Temporal Kohonen Map [6], recurrent SOM [7], feedback SOM [8], recursive SOM [9], merge SOM [10], SOM for structured data [11] and contextual SOM for structured data [12].

Nevertheless, the heuristic nature of SOM inherently brings about certain limitations, for example the lack of a principled cost function (however in [13] a slightly modified version of SOM is presented that does possess a well-defined cost function). Comparison of topographic maps that resulted from different initializations, parameter settings, or optimization algorithms can be problematic. Moreover extensions of SOM, such as the ones mentioned above, also inherit the problems related to the absence of a cost function (although see developments in [2] along the lines of [13]). Moreover, it is difficult to understand what is driving the learning of topographic maps since it is not clear what the optimization objective is. Clusters formed on the map can indicate some affinity between the concerned structured data items, but mere inspection of the learned map does not explain the driving forces behind its formation. In particular, reasoning about mapping of new data items (not used during training) can be challenging.

The Generative Topographic Mapping (GTM) algorithm [14] constructs topographic maps of vectorial data. It is based on sound probabilistic foundations. The GTM is a constrained mixture of local noise models (spherical Gaussian densities). The mixture of Gaussians is constrained in the sense that the means of the Gaussians are forced to reside on a low dimensional manifold in a high dimensional data space. Training the model can then be understood as fitting (in the distribution sense) the manifold to the training data by adjusting the model parameters.

In this work, we present extensions of GTM to the visualization of sequences and tree-structured data. These extensions are principled probabilistic model-based formulations which enjoy several benefits over recursive neural-based approaches. One of the most important advantages is that a cost function arises naturally as the likelihood of the model. Optimization of this cost function constitutes the driving force in learning topographic maps. Also the cost function allows the detection of overfitting by monitoring its evolution on an independent dataset (not used in training). The generative nature of the presented model explains how the data items might have been generated. This is valuable in understanding map formations and why data items are projected to their particular locations on the topographic map.

2   The Generative Topographic Mapping Algorithm

Consider a dataset of d-dimensional vectors, $\mathcal{T} = \{t^{(1)}, \ldots, t^{(N)}\}$, assumed to have been independently generated along a smooth q-dimensional manifold embedded in $\mathbb{R}^d$. Such a dataset can be modeled by a mixture of C spherical Gaussians:

$$p(\mathcal{T}) = \prod_{n=1}^{N} \sum_{m=1}^{M} P(c)p(t^{(n)}|c) = \prod_{n=1}^{N} \sum_{c=1}^{C} P(c)\mathcal{N}(t^{(n)}; \boldsymbol{\mu}_c, \sigma_c), \qquad (1)$$

where $P(c)$ are the mixing coefficients, $\boldsymbol{\mu}_c$ the means of the Gaussians and σ_c the standard deviations. We assume $P(c) = \frac{1}{C}$ and fixed variance $\sigma_c^2 = \sigma^2$. This model is an unconstrained model in the sense that its parameters, the means, do not adhere to any constraints and can move freely. In order to capture topographic organization of the data points, we impose that the means $\boldsymbol{\mu}$ lie on an image of a q-dimensional interval (latent space) $\mathcal{V}$ ($q < d$, $q = 2$ for the purposes of visualization) under a smooth mapping $\Gamma : \mathcal{V} \to \mathbb{R}^d$. The non-linear mapping Γ takes the form [14]:

$$\Gamma(\boldsymbol{x}) = \boldsymbol{W}\boldsymbol{\phi}(\boldsymbol{x}), \qquad (2)$$

which is a RBF network with basis functions $\boldsymbol{\phi}(\cdot)$ and weight matrix $\boldsymbol{W}$. Function Γ maps each latent point $\boldsymbol{x} \in \mathcal{V}$ to the local noise model mean $\boldsymbol{\mu}$ in a non-linear manner. Since Γ is smooth, the projected points retain their local neighborhood in the space $\mathbb{R}^d$. Thus, a topographic organization is imposed on the means $\boldsymbol{\mu}$ of the Gaussians. We discretize the space $\mathcal{V}$ by a rectangular grid of points $\boldsymbol{x}_c$, $c = 1, \ldots, C$ and impose a prior distribution on the latent space:

$$p(\boldsymbol{x}) = \frac{1}{C} \sum_{c=1}^{C} \delta(\boldsymbol{x}_c - \boldsymbol{x}), \qquad (3)$$

where $\delta(\boldsymbol{x})$ denotes the Dirac delta function ($\delta(\boldsymbol{x}) = 1$ for $\boldsymbol{x} = 0$, otherwise $\delta(\boldsymbol{x}) = 0$). The likelihood function reads:

$$\mathcal{L} = p(\mathcal{T}) = \frac{1}{C} \prod_{n=1}^{N} \sum_{c=1}^{C} \mathcal{N}(t^{(n)}; \boldsymbol{\mu}_c, \sigma), \qquad (4)$$

where $\boldsymbol{\mu}_c = \Gamma(\boldsymbol{x}_c)$.

The Expectation-Maximization (EM) algorithm [14] is employed to optimize the likelihood $\mathcal{L}$. Once the model is trained, we can visualize the dataset using the posteriors

$$p(\boldsymbol{x}_c|\boldsymbol{t}^{(n)}) = \frac{p(\boldsymbol{t}^{(n)}|\boldsymbol{x}_c)P(\boldsymbol{x}_c)}{p(\boldsymbol{t}^{(n)})} = \frac{\mathcal{N}(\boldsymbol{t}^{(n)};\boldsymbol{\mu}_{\boldsymbol{x}_c},\sigma)}{\sum_{c'=1}^{C}\mathcal{N}(\boldsymbol{t}^{(n)};\boldsymbol{\mu}_{\boldsymbol{x}_{c'}},\sigma)}. \tag{5}$$

Data item $\boldsymbol{t}^{(n)}$ is then represented on the latent space by the mean of the posterior distribution over the latent centers points $\boldsymbol{x}_c$:

$$proj(\boldsymbol{t}^{(n)}) = \sum_{c=1}^{C}p(\boldsymbol{x}_c|\boldsymbol{t}^{(n)})\boldsymbol{x}_c. \tag{6}$$

3   Overview of Hidden Markov Tree Models

As we saw in the GTM, the building component is the spherical Gaussian density. Each latent point $\boldsymbol{x} \in \mathcal{V}$ is mapped via Γ to a point in $\mathbb{R}^d$ which is taken to be the mean of a spherical Gaussian of a fixed variance σ^2 and so we associate each $\boldsymbol{x} \in \mathcal{V}$ with a Gaussian density $p(\cdot|\boldsymbol{x})$. However, when working with sequences and trees Gaussians are inappropriate as noise models and need be replaced by a more suitable probabilistic model; in the case of sequences our substitution is the hidden Markov model (HMM) [15] and in the case of tree-structures we choose hidden Markov trees models (HMTM). Since HMMs are well-known due to their ubiquitous applications, we present a brief overview only on the closely related HMTMs.

A tree $\boldsymbol{y}$ is an acyclic directed graph that consists of a set of nodes $u \in \mathcal{U}_{\boldsymbol{y}} = \{1, 2, ..., U_{\boldsymbol{y}}\}$, a set of directed edges between parent and children nodes and a set of labels $\boldsymbol{o}_u \in \mathbb{R}^d$ on the nodes. Each node u has a single parent $\rho(u)$ (apart from the root node) and each node u has a set of children $ch(u)$ (apart from the leaf nodes). Moreover, we introduce notation for subtrees. A subtree rooted at node u of a tree $\boldsymbol{y}$ is denoted by $\boldsymbol{y}_u$. The entire tree $\boldsymbol{y}$ is identified with the subtree $\boldsymbol{y}_1$. Also, $\boldsymbol{y}_{u \backslash v}$ denotes the entire subtree $\boldsymbol{y}_u$, excluding the subtree rooted at node v. A notated tree is depicted in Fig. 1(a).

A hidden state process is defined that labels the nodes of the trees. This process associates with each node u a discrete random variable Q_u which can enter one of K unobservable states. Variable Q_u stochastically determines the label for node u. Each state $k = 1, 2, ..., K$ is associated with a parametrized emission distribution $f(.; \boldsymbol{\psi}_k)$ that produces a label. So given a tree structure $\boldsymbol{y}$, the model assigns each node u a label $\boldsymbol{o}_u$ using $f(.; \boldsymbol{\psi}_k)$ which depends on the state $Q_u \in \{1, 2, ..., K\}$ of node u. The state Q_u of each node u is stochastically determined by a joint probability distribution over state variables. According to this joint distribution each node u is dependent on its parent $\rho(u)$. Thus, the state Q_u is conditioned on the state $Q_{\rho(u)}$ of its parent. Since the state process is hidden, what we observe is the result of the process i.e. the labels $\boldsymbol{o}_u$ of the nodes u in the tree, but the hidden states Q_u entered by nodes u are not observed. This is illustrated in Fig. 1(b). This model is known as the *hidden Markov tree model* (HMTM) [17,18], and is an extension of the HMM.

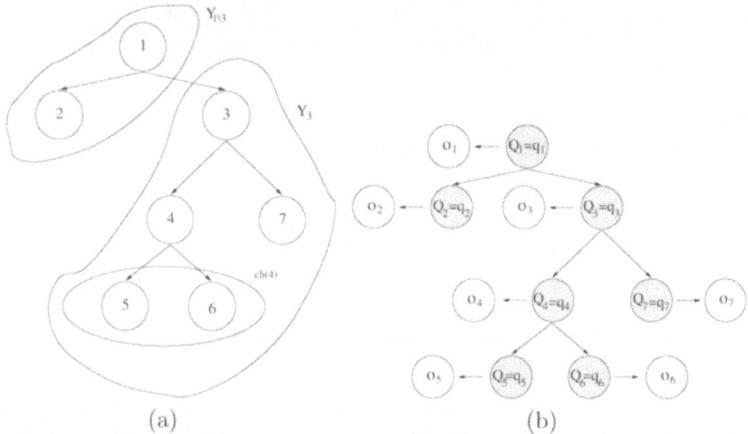

Fig. 1. Notation in tree structures (a), Example of an underlying hidden state (states in gray) process emitting labels (b) ([16] ©IEEE 2008)

A HMTM is defined by three sets of parameters:

- initial probability vector $\boldsymbol{\pi} = \{p(Q_1 = k)\}_{k=1,\ldots,K}$ – each element expressing the probability of the root node being in state $k \in \{1, 2, \ldots, K\}$.
- transition probability matrix $\boldsymbol{B} = \{p(Q_u = l|Q_{\rho(u)} = k)\}_{k,l=1,\ldots,K}$ – each element expressing the probability of transiting from parent $\rho(u)$ in state k to the child node u in state l. This probability is assumed to be position-invariant. Note that the sum of elements b_{kl} over $l = 1, \ldots, K$, i.e. the sum of transitions from a state k to all other K states, must equal 1.
- the emission parameters that parametrize Gaussian distributions, $f(.; \boldsymbol{\psi}_k)$, with $\boldsymbol{\psi}_k = \{\boldsymbol{\mu}_k, \boldsymbol{\Sigma}_k\}$ one for each state $k = 1, \ldots, K$. Here, $\boldsymbol{\mu}_k \in \mathbb{R}^d$ and $\boldsymbol{\Sigma}_k$ are the mean and covariance matrix, respectively, of the Gaussian associated with emission process in state k.

The Markovian dependencies of hidden states are realized by the following conditions [18]:

- Given the parent state $Q_{\rho(u)}$, the child state Q_u is conditionally independent of all other variables in the tree, apart from those that belong in the subtree $\boldsymbol{y}_u$:

$$p(Q_u = q_u|\{Q_v = q_v\}_{v \in \mathcal{U}_{\boldsymbol{y}}, v \notin \boldsymbol{y}_u}, \{O_v = o_v\}_{v \in \mathcal{U}_{\boldsymbol{y}}, v \notin \boldsymbol{y}_u}) =$$
$$p(Q_u = q_u|Q_{\rho(u)} = q_{\rho(u)}). \qquad (7)$$

- Given the (hidden) state of a node, the corresponding label is conditionally independent of all other variables in the tree:

$$p(O_u = o_u|\{O_v = o_v\}_{v \in \mathcal{U}_{\boldsymbol{y}}, v \neq u}, \{Q_v = q_v\}_{v \in \mathcal{U}_{\boldsymbol{y}}}) = p(O_u = o_u|Q_u = q_u). \qquad (8)$$

Thus, a HMTM expresses a joint distribution that can be factorized as follows:

$$p(\boldsymbol{y}, Q_1 = q_1, \ldots, Q_{U_{\boldsymbol{y}}} = q_{U_{\boldsymbol{y}}}) = p(Q_1 = q_1) \prod_{u \in \mathcal{U}_{\boldsymbol{y}}, u \neq 1} p(Q_u = q_u | Q_{\rho(u)} = q_{\rho(u)})$$

$$\times \prod_{u \in \mathcal{U}_{\boldsymbol{y}}} p(\boldsymbol{O}_u = \boldsymbol{o}_u | Q_u = q_u). \tag{9}$$

In order to ease the notation, in the text following we shall not state both random variables and their instantiations, but only the latter: $p(\boldsymbol{O}_u = \boldsymbol{o}_u | Q_u = q_u) = p(\boldsymbol{o}_u | q_u)$. Inference in HMTMs is conducted via the upward-downward algorithm [19] which is the direct analogue of the forward-backward algorithm in HMMs.

4   Extending the GTM to Sequences and Tree-Structured Data

One may view the GTM as consisting of two components, a generative probabilistic model, and a suitable constrained parametrization. For a given data type, there can be many different choices of local noise models in the data space. Since in the GTM formulation, two data items are viewed as being "close" if they are generated by the same (or similar) local noise model, the nature of the topographic map is determined by our choice of noise model. The methodology is general and can easily accommodate different notions of similarity between structured data items.

4.1   Hidden Markov Trees as Noise Models for the GTM

In this Sect. we extend the GTM to the visualization of tree-structured data. Here the observations are no longer fixed-length vectors $\boldsymbol{t}$, but trees $\boldsymbol{y}$, as described in Sect. 3. Hence analogously to GTM, each latent point $\boldsymbol{x} \in \mathcal{V}$ is mapped to an HMTM via a smooth non-linear mapping Γ. Since the neighborhood of Γ-images of $\boldsymbol{x}$ is preserved, the resulting HMTMs will be topographically organized. Likelihoods $p(\boldsymbol{y}|\boldsymbol{x})$ are calculated using the upward-downward algorithm [19]. Again we commence the formulation of the model based on the form of a standard mixture model. Assuming that the given trees $\mathcal{T} = \{\boldsymbol{y}^{(1)}, \boldsymbol{y}^{(2)}, ..., \boldsymbol{y}^{(N)}\}$ are independently generated, the model likelihood is:

$$\mathcal{L} = \prod_{n=1}^{N} p(\boldsymbol{y}^{(n)}) = \prod_{n=1}^{N} \sum_{c=1}^{C} p(\boldsymbol{y}^{(n)} | \boldsymbol{x}_c) p(\boldsymbol{x}_c), \tag{10}$$

where the mixing coefficients are ignored since $p(\boldsymbol{x}) = \frac{1}{C}$. The number of nodes $U_{\boldsymbol{y}^{(n)}}$ in n-th tree $\boldsymbol{y}^{(n)}$ is denoted by U_n. We consider a subset of points in the latent space $\mathcal{V}$ that form a regular grid $\{\boldsymbol{x}_c\}_{c=1}^{C}$. Noise models $p(\boldsymbol{y}^{(n)} | \boldsymbol{x}_c)$ are expanded using (9):

$$\mathcal{L} \propto \prod_{n=1}^{N} \sum_{c=1}^{C} \sum_{\boldsymbol{q} \in \{1,2,...,K\}^{U_n}} p(q_1|\boldsymbol{x}_c) \prod_{u=2}^{U_n} p(q_u|q_{\rho(u)}, \boldsymbol{x}_c) \prod_{u=1}^{U_n} p(\boldsymbol{o}_u^{(n)}|q_u, \boldsymbol{x}_c). \quad (11)$$

In order to have the HMTM components topologically organized — e.g. on a two-dimensional equidistant grid — we constrain the mixture of HMTMs,

$$p(\boldsymbol{y}) = \frac{1}{C} \sum_{c=1}^{C} p(\boldsymbol{y}|\boldsymbol{x}_c), \quad (12)$$

by requiring that the HMTM parameters be generated through a parametrized *smooth* nonlinear mapping from the latent space into the HMTM parameter space:

$$\boldsymbol{\pi}_c = \{p(q_1 = k|\boldsymbol{x}_c)\}_{k=1,...,K} \{g_k(\boldsymbol{A}^{(\boldsymbol{\pi})} \boldsymbol{\phi}(\boldsymbol{x}_c))\}_{k=1,...,K}, \quad (13)$$

$$\boldsymbol{T}_c = \{p(q_u = k|q_{\rho(u)} = l, \boldsymbol{x}_c)\}_{k,l=1,...,K} \{g_k(\boldsymbol{A}^{(\boldsymbol{T}_l)} \boldsymbol{\phi}(\boldsymbol{x}_c))\}_{k,l=1,...,K}, \quad (14)$$

$$\boldsymbol{B}_c = \{\boldsymbol{\mu}_k^{(c)}\}_{k=1,...,K} \{\boldsymbol{A}^{(\boldsymbol{B}_k)} \boldsymbol{\phi}(\boldsymbol{x}_c)\}_{k=1,...,K}, \quad (15)$$

where

- the function $g(\cdot)$ is the softmax function, which is the canonical inverse link function of multinomial distribution and $g_k(\cdot)$ denotes the k-th component returned by the softmax, i.e.

$$g_k\left((a_1, a_2, ..., a_q)^T\right) = \frac{e^{a_k}}{\sum_{i=1}^{q} e^{a_i}}, \quad k = 1, 2, ..., q.$$

Here the softmax function "squashes" the values of $\boldsymbol{A}^{(\boldsymbol{\pi})} \boldsymbol{\phi}(\boldsymbol{x}_c)$ and $\boldsymbol{A}^{(\boldsymbol{T}_l)} \boldsymbol{\phi}(\boldsymbol{x}_c)$, which are unbounded, to values in the range $[0, 1]$. This is necessary as the elements in $\boldsymbol{\pi}_c$ and $\boldsymbol{T}_c$ are probabilities that are naturally constrained in $[0, 1]$.
- $\boldsymbol{x}_c \in \mathbb{R}^2$ is the c-th grid point in the latent space $\mathcal{V}$,
- $\boldsymbol{\phi}(\cdot) = (\phi_1(\cdot), ..., \phi_M(\cdot))^T, \phi_m(\cdot): \mathbb{R}^2 \to \mathbb{R}$ a vector function consisting of M nonlinear smooth basis functions (typically RBFs),
- the matrices $\boldsymbol{A}^{(\boldsymbol{\pi})} \in \mathbb{R}^{K \times M}$, $\boldsymbol{A}^{(\boldsymbol{T}_l)} \in \mathbb{R}^{K \times M}$ and $\boldsymbol{A}^{(\boldsymbol{B}_k)} \in \mathbb{R}^{d \times M}$ are the free parameters of the model.

We employ the EM algorithm to maximize the model likelihood. For derivation and implementation details, see [16].

Once the model has been trained, each tree data item $\boldsymbol{y}^{(n)}$ is mapped to a point $proj(\boldsymbol{y}^{(n)})$ on the latent space that is given by the mean of the posterior distribution over all latent points $\boldsymbol{x}$:

$$proj(\boldsymbol{y}^{(n)}) = \sum_{c=1}^{C} p(\boldsymbol{x}_c|\boldsymbol{y}^{(n)}) \boldsymbol{x}_c. \quad (16)$$

4.2   Hidden Markov Models as a Noise Model for GTM

Having presented the extension of the GTM to constructing topographic maps of tree-structured data, the extension to sequences can be viewed as a special case. One may view sequences as a degenerate form of trees where each node in the tree is allowed no more than one child node. In [20] an extension of the GTM is presented, which we term GTM-HMM, that constructs topographic maps of sequences. The model formulation of the GTM-HMM is very close to that of GTM-HMTM. However, we note that in [20] the sequences considered are discrete with the symbols o_t belonging to a fixed finite alphabet $\mathcal{S} = \{1, 2, \ldots, S\}$. Therefore, in [20] even though the initial probabilities and transition are parametrized in the same way as shown in (13) and (14), emission parameters are parametrized using:

$$\boldsymbol{B}_c = \{p(q_t = k | q_{t-1} = k, \boldsymbol{x}_c)\}_{s=1,\ldots,S,\ k=1,\ldots,K}$$
$$= \{g_s(\boldsymbol{A}^{(\boldsymbol{T}_k)} \phi(\boldsymbol{x}_c))\}_{s=1,\ldots,S,\ k=1,\ldots,K}. \tag{17}$$

The GTM-HMM extension is trained using the EM algorithm with the E- and M-steps being almost identical those of the GTM-HMTM in Sect. 4.1.

5   Magnification Factors

In the standard GTM, topographic organization of the data on the two-dimensional latent space $\mathcal{V}$ allows the inspection of spatial data relationships in the high dimensional space $\mathbb{R}^d$ and the inference of potential clusters and relations between the data points $\boldsymbol{t}$. However, we must keep in mind that the data points are projected on the latent space in a non-linear way. This means that data points that are distant in the data space (assuming some notion of a metric in the data space) may be projected close to each other. Thus, even though the smooth mapping does preserve the neighborhood structure, it does not necessarily preserve distances in the latent space.

Magnification factors for the GTM are introduced in [21]. Each point $\boldsymbol{x}$ of latent space $\mathcal{V}$ is mapped to the mean $\boldsymbol{\mu} \in \mathbf{R}^d$ of a spherical Gaussian density via $\Gamma(\boldsymbol{x}) = \boldsymbol{\mu} = \boldsymbol{W}\phi(\boldsymbol{x})$. Thus, assuming a fixed variance σ^2 of the spherical Gaussians (standard GTM model), a smooth two-dimensional statistical manifold of spherical Gaussians is induced in the space of all spherical Gaussians of variance σ^2. We denote this induced manifold by $\mathcal{M}$. In the case of standard GTM we can identify $\mathcal{M}$ with $\Gamma(\mathcal{V})$.

In [21] an infinitesimal rectangle is defined at a point $\boldsymbol{x} \in \mathcal{V}$ by taking displacements $d\boldsymbol{x}_1 = dx_1 \boldsymbol{e}_1$, $d\boldsymbol{x}_2 = dx_2 \boldsymbol{e}_2$ along a Cartesian coordinate system $\{\boldsymbol{e}_1, \boldsymbol{e}_2\}$ in $\mathcal{V}$. The area of this rectangle is equal to $dA = dx_1 dx_2$. Also, the area of the Γ-image of this rectangle is $dA' = dA \cdot \det(\boldsymbol{J}^T \boldsymbol{J})$ [21]. where $\boldsymbol{J}$ is the Jacobian of mapping Γ at $\boldsymbol{x}$. The magnification factor at point $\boldsymbol{x} \in \mathcal{V}$ is calculated at the ratio of the two areas: $\frac{dA'}{dA} = \det(\boldsymbol{J}^T \boldsymbol{J})$.

In this work we present two alternative methods for appreciating magnification factors around latent points $\boldsymbol{x} \in \mathcal{V}$ that encompass the aforementioned method used in [21], and are also applicable on the presented extensions of the GTM. Both methods rely on the concept of Kullback-Leibler divergence. The Kullback-Leibler divergence (KLD) is a scalar quantity that informs us of the "distance" between two distributions P and Q (assuming that $Q(t) = 0 \Rightarrow P(t) = 0$) and is defined as [22]:

$$D_{KL}[P||Q] = \int P(t) \log \frac{P(t)}{Q(t)} dt. \tag{18}$$

and for the discrete case:

$$D_{KL}[P||Q] = \sum_i P_i \log \frac{P_i}{Q_i}, \tag{19}$$

where P_i and Q_i are probabilities of event i under P and Q, respectively.

An important property, known as Gibbs' inequality, is that KLD is always non-negative with a minimum of zero when Q exactly matches distribution P, $D_{KL}[P||P] = 0$ [22]. It is important to note that KLD is not symmetric in general $D_{KL}[P||Q] \neq D_{KL}[Q||P]$, hence it does not constitute a proper distance metric but can be interpreted as a *statistical distance* of approximating distribution P with distribution Q. In practice, for two distributions P and Q KLD can be measured as the observed $\hat{D}_{KL}$ via a Monte-Carlo type approximation:

$$\hat{D}_{KL}[P||Q] = \sum_{t \in \mathcal{T}} P(t) \log P(t) - \sum_{t \in \mathcal{T}} P(t) \log Q(t).$$

Provided the distributions are smoothly parametrized, the KLD between two distributions with close parameter settings can be approximated via *Fisher information matrix* (FIM) acting as a metric tensor. This concept makes explicit use of the geometry of the models in the manifold $\mathcal{M}$. Note that our extensions of GTM are smoothly parametrized. Each latent point has two coordinates $\boldsymbol{x} = (x_1, x_2)^T \in \mathcal{V}$ and is mapped via a smooth non-linear mapping to a noise model $p(\cdot|\boldsymbol{x})$ on the manifold $\mathcal{M}$ of HMTMs. If we displace $\boldsymbol{x}$ by an infinitesimally small perturbation $d\boldsymbol{x}$, the KLD $D_{KL}[p(\cdot|\boldsymbol{x})||p(\cdot|\boldsymbol{x} + d\boldsymbol{x})]$ between the corresponding noise models $p(\cdot|\boldsymbol{x}), p(\cdot|\boldsymbol{x} + d\boldsymbol{x}) \in \mathcal{M}$ can be approximated via Fisher information matrix

$$\boldsymbol{F}(\boldsymbol{x}) = -E_{p(\cdot|\boldsymbol{x})}[\nabla^2 \log p(.|\boldsymbol{x})], \tag{20}$$

that acts like a metric tensor on the Riemannian manifold $\mathcal{M}$ [23]:

$$D_{KL}[p(\cdot|\boldsymbol{x})||p(\cdot|\boldsymbol{x} + d\boldsymbol{x})] = d\boldsymbol{x}^T \boldsymbol{F}(\boldsymbol{x}) \, d\boldsymbol{x}. \tag{21}$$

The situation is illustrated in Fig. 2. In our experiments, each latent point $\boldsymbol{x}$ is perturbed systematically in various directions by a small displacement $d\boldsymbol{x}$ as illustrated in Fig. 3.

In the case of GTM-HMTM, we base the calculation of the observed Fisher information matrix on the upward recursion of the likelihood estimation for HMTMs [17]. In the case of GTM-HMM, the same procedure is applied, treating sequences as degenerate trees with out-degree 1. For detailed derivations, see [16,24].

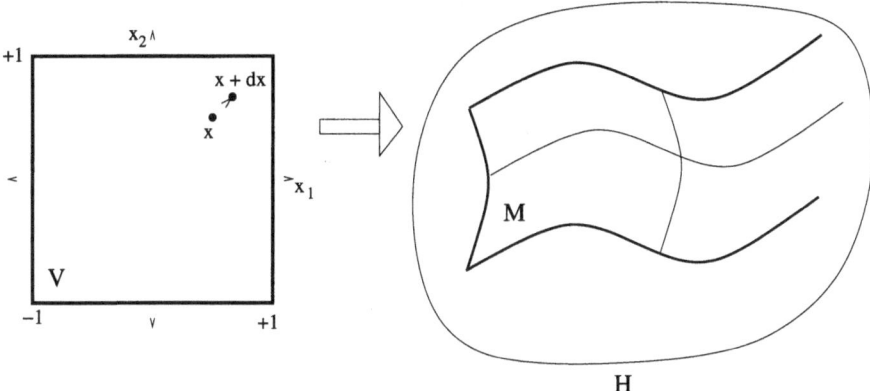

Fig. 2. Two-dimensional manifold $\mathcal{M}$ of local noise models $p(\cdot|\boldsymbol{x})$ parametrized by the latent space $\mathcal{V}$ through (2). The manifold is embedded in manifold $\mathcal{H}$ of all noise models of the same form. Latent coordinates $\boldsymbol{x}$ are displaced to $\boldsymbol{x} + \boldsymbol{dx}$. Kullback-Leibler divergence $D_{KL}[p(\cdot|\boldsymbol{x})\|p(\cdot|\boldsymbol{x} + \boldsymbol{dx})]$ between the corresponding noise models $p(\cdot|\boldsymbol{x}), p(\cdot|\boldsymbol{x} + \boldsymbol{dx}) \in \mathcal{M}$ can be determined via Fisher information matrix $\boldsymbol{F}(\boldsymbol{x})$ that acts like a metric tensor on the Riemannian manifold $\mathcal{M}$.

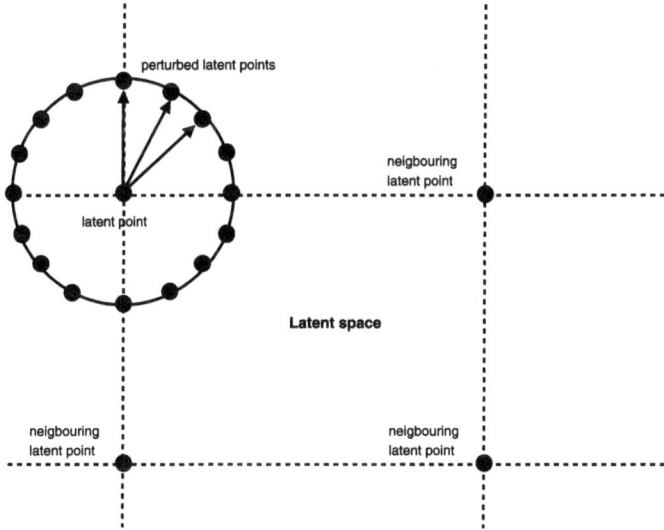

Fig. 3. Magnification factors may be measured via the perturbation of a latent point in regular intervals on a small circle

In order to illustrate the magnification factors on manifold $\mathcal{M}$, we calculate the observed FIM for each latent center $\boldsymbol{x}_c, c = 1, 2, \ldots, C$. We can then compute the observed KLD between each $p(\cdot|\boldsymbol{x}_c)$ and its perturbation $p(\cdot|\boldsymbol{x}_c + \boldsymbol{dx})$. We perturbed each latent center $\boldsymbol{x}_c$ in 16 regularly spaced directions on a small circle (we have set its radius to 10^{-5}). This is illustrated in Fig. 3 for 16 directions. We note that alternatively, we could have used SVD decomposition of the FIM to find and quantify the local dominant stretching directions in the latent space.

We have also developed an alternative fast approximation scheme for calculating the observed FIM of HMTMs based on findings in [25]. The approximation is based on the upward recursion [17] and provides an upper bound for KLD. As in KLD calculations involving Fisher information matrix, we perturb each latent center in 16 regularly spaced directions on a small circle (again the radius is set to 10^{-5}). We then estimate the KLD between the original HMTM model x and the perturbed model $x + dx$.

6   Experimental Results

In all experiments reported below, the lattice in the latent space $\mathcal{V} = [-1, 1]^2$ was a 10×10 regular grid (i.e. $C = 100$) and the RBF network consisted of $M = 17$ basis functions; 16 of them were Gaussian radial basis functions of variance 1.0 centered on a 4×4 regular grid, and one was a constant function $\phi_{17}(x_c) = 1$ intended to serve as a bias term. Parameters were initialized randomly with uniform distribution in $[-1, 1]$. In GTM-HMTM, the state-conditional emission probability distributions were modeled as multivariate spherical Gaussians.

6.1   Artificially Generated Sequences

We first generated a toy data set of 400 binary sequences of length 40 from four HMMs ($K = 2$ hidden states) with identical emission structure, i.e. the HMMs differed only in transition probabilities. Each of the four HMMs generated 100 sequences. Local noise models $p(\cdot|x)$ were imposed as HMMs with 2 hidden states. Visualization of the sequences is presented in Fig. 4. Sequences are marked with four different markers, corresponding to the four different HMMs used to generate the data set. We stress that the model was trained in a completely unsupervised way. The markers are used for illustrative purposes only. Representations of induced metric in the local noise model space based on Fisher information matrix and direct fast KLD estimation can be seen in Fig. 5(a) and Fig. 5(b), respectively. Dark areas signify homogeneous regions of local noise models and correspond to possible clusters in the data space. Light areas signify abrupt changes in local noise model distributions (as measured by KLD) and correspond to boundaries between data clusters. The visualization plot reveals that the latent trait model essentially discovered the organization of the data set and the user would be able to detect the four clusters, even without help of the marking scheme in Fig. 4. Of course, the latent trait model benefited from the fact that the distributions used to generate data were from the same model class as the local noise models.

6.2   Melodic Lines of Chorals by J.S.Bach

In this experiment we visualize a set of 100 chorals by J.S. Bach [26]. We extracted the melodic lines – pitches are represented in the space of one octave,

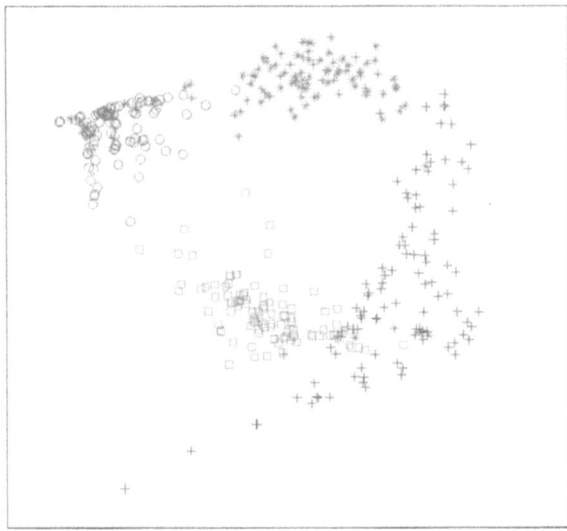

Fig. 4. Toy sequences: visualization of 400 binary sequences of length 40 generated by 4 HMMs with 2 hidden states and with identical emission structure. Sequences are marked with four different markers, corresponding to the 4 HMMs.

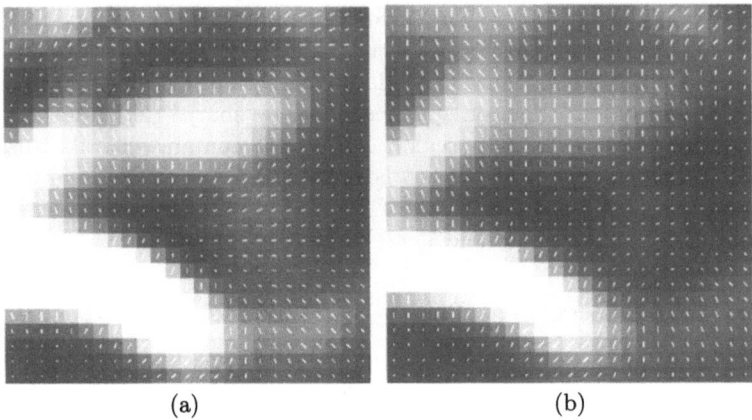

(a)   (b)

Fig. 5. Toy sequences: magnification factors via FIM (a) and KLD (b) for GTM-HMM on toy dataset of sequences

i.e. the observation symbol space consists of 12 different pitch values. Temporal structure of the sequences is the essential feature to be considered when organizing the data items in any sensible manner. Local noise models had $K = 3$ hidden states.

Figure 6 shows choral visualizations, while representations of induced metric in the local noise model space based on Fisher Information matrix and direct

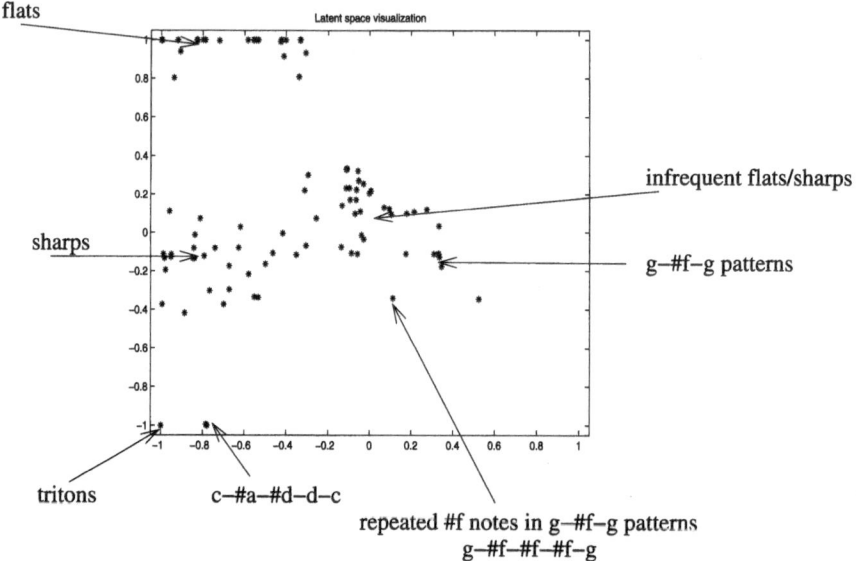

Fig. 6. Visualization of melodic lines of 100 chorals by J.S. Bach

K-L divergence estimations can be seen in figures 7(a) and 7, respectively. The method discovered natural topography of the key signatures, corroborated with similarities of melodic motives. The melodies can contain sharps and flats other than those included in their key signature due to both modulation and ornaments. The upper region contains melodic lines that utilize keys with flats. Central part of the visualization space is taken by sharps (left) and almost no sharps/flats (center). The lower region of the plot contains chorals with tense patterns (e.g containing tritons) and is quite clearly strongly separated from other chorals. The overall clustering of chorals is well-matched by the metric representations of figures 7(a) and 7.

6.3   Artificially Generated Trees

The dataset is produced by sampling from 4 HMTMs with $K = 2$ hidden states and two-dimensional Gaussian emissions of fixed spherical variance, each corresponding to one of the four classes (HMTMs). Each artificial class was populated by 80 tree samples. All 320 sampled trees possess the same fixed topology of a binary tree of 15 nodes. Moreover, we ensured that the parameters of the 4 HMTMs were such that it would be impossible to determine the class of a sample by simply looking only at its labels and ignoring its structure. The parameters that were used to parametrize the 4 HMTMs are summarized in Table 1.

Fig. 8 presents the topographic map constructed by GTM-HMTM for the toy dataset (the number of hidden states was set to $K = 2$). Each sampled tree is represented by one of four different markers that signify class

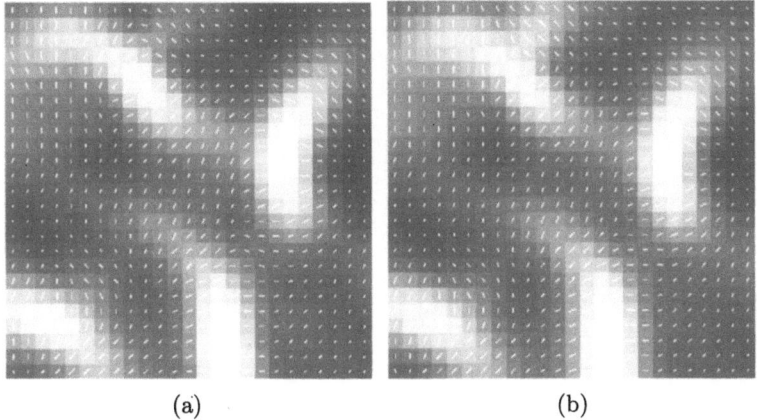

(a)   (b)

Fig. 7. Chorals task: magnification factors via FIM (a) and KLD (b) for GTM-HMM on Bach chorals dataset

Table 1. Parameters of HMTMs for creating the toy dataset. Variance was fixed to $\sigma^2 = 1$.

Class	Initial prob	Transition prob	Means of emissions
HMTM 1	0.7 0.3	0.9 0.1 0.1 0.9	−1.0 4.0 1.0 2.0
HMTM 2	0.7 0.3	0.9 0.1 0.1 0.9	−2.0 6.0 3.0 0.0
HMTM 3	0.7 0.3	0.1 0.9 0.9 0.1	−1.0 4.0 1.0 2.0
HMTM 4	0.7 0.3	0.1 0.9 0.9 0.1	−2.0 6.0 3.0 0.0

membership to one of the four generative classes used to construct the data set. The separation of the classes into 4 distinct clusters clearly displays the high level of topographic organization that the model has been achieved. Regarding the covariance of the emission distribution, we initialized it to $\Sigma_k = 2I$ for both states $k = 1, 2$ where I stands for the identity matrix. However, initializing it with $\Sigma_k = 2I, 3I, 5I$ also lead to maps of similar level of topographic organization.

We also calculated magnifications factors using both approaches of FIM and local KLD approximation which yielded identical results. The magnification factors are illustrated in Fig. 9. The magnification factors indicate that the 4 clusters are well separated by bright regions. We observe that sample trees of common class origin are clustered together, while sample trees belonging to different classes have greater separations between them.

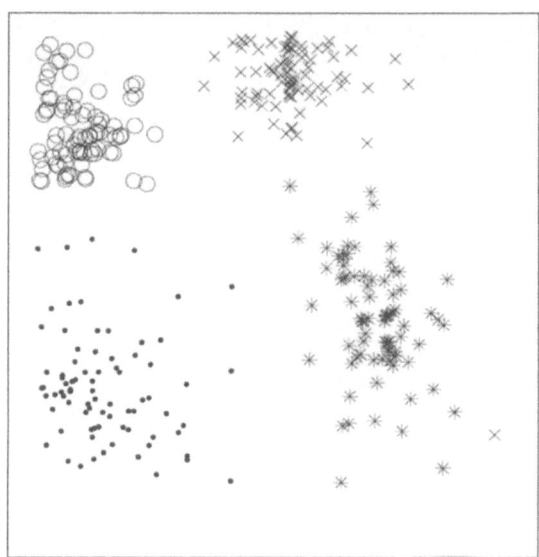

Fig. 8. Toy trees: visualization of toy dataset using GTM-HMTM ([16] ©IEEE 2008)

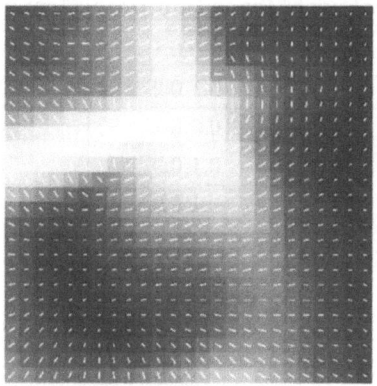

Fig. 9. Toy trees: visualization of toy dataset using GTM-HMTM ([16] ©IEEE 2008)

6.4   Traffic Policeman Benchmark Tree Structured Data

The second dataset of tree-structured data consists of benchmark images produced by the *Traffic Policeman Benchmark* (TPB) software [27]. This software
provides a testbed for evaluating learning algorithms that process structured patterns. The images produced by the software resemble traffic policemen, houses
and ships of different shape, size and color. Three sample images of each type
are illustrated in the in Fig. 10(a), 10(b) and 10(c). Connected components (i.e.
components that are adjacent and touch) in each image have a parent-child relationship, the object located lower and closer to the left edge being the parent

Table 2. Classes in TPB dataset

Class	Symbol	Description
A	○	Policemen with the lowered left arm
B	x	Policeman with the raised left arm
C	*	Ships with two masts
D	•	Ships with three masts
E	△	Houses with one upper right window
F	▽	Houses with upper left and lower left window
G	◁	Houses with two upper windows
H	▷	Houses with lower left and upper right window
I	★	Houses with three windows
J	□	Houses with one lower left window
K	+	Houses with no windows
L	◇	Houses with one upper left window

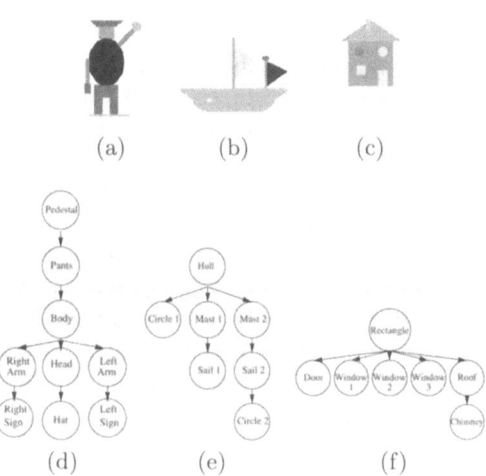

(a)   (b)   (c)

(d)   (e)   (f)

Fig. 10. Sample images from TPB in (a), (b), (c) and their corresponding tree representations in (d), (e), (f) ([16] ©IEEE 2008)

(i.e. the images must be interpreted bottom-up, left to right). Fig. 10(d), 10(e) and 10(f) display the tree representations of the images corresponding to Fig. 10(a), 10(b) and 10(c). Actually the TPB software produces general acyclic graph structures, but here we restricted it to produce strictly images that have a tree representation. Each node in each tree is labeled with a two-dimensional vector. This two-dimensional vector is a pair of coordinates for the center of gravity of the component that node stands for. The dataset consists 12 classes, each populated by 50 samples. Table 2 lists the classes of the dataset.

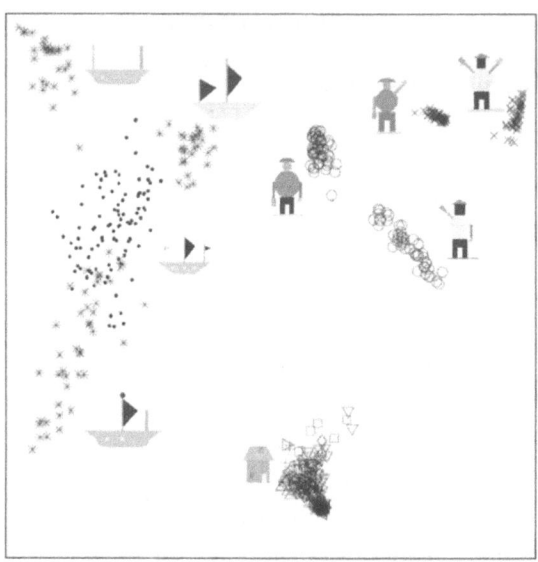

Fig. 11. TPB: visualization of TPB dataset using GTM-HMTM ([16] ©IEEE 2008)

Fig. 11 presents the topographic map constructed by GTM-HMTM on the TPB dataset. Next to each cluster a representative image is displayed. Judging alone from the markers, it is evident that the model has achieved a level of topographic organization. It is interesting to note the emerging sub-classes within the defined classes of Table 2. Class × has been split into two sub-clusters, one with policemen with the right arm lowered and one with the right arm raised. Also, class ◯ has been divided into policemen with the right arm lowered and policemen with the arm raised. An interesting behavior is exhibited by the sub-clusters of ships as not only has class ∗ been divided into three sub-clusters, but the sub-clusters that surround class • which possibly indicates how the ships classes are related. Thus, class • seems to link the three discovered sub-clusters of ships; class • represents ships with three masts, while the three sub-clusters around class ∗ are composed of ships with either the two masts, with either the left, center or right mast missing. Nevertheless, the model has not been successful in the visualization of the classes representing houses. No distinct clusters of houses have been formed. Instead all house structures have been organized into a single cluster that represents the images of all houses. One possible explanation for this inability of discriminating between the classes of houses, is the shallow tree representation of houses; typically they are shorter than ships and traffic-policemen structures.

7   Discussion

The clear model-based formulation of the GTM extensions permits a clear understanding of the topographic organization in the induced maps. For example in the case of GTM-HMTM, we inspected in detail the parameters of the underlying local hidden Markov tree models and provided explanations of why data items are projected to their particular locations on the map [16]. The type of noise model employed by the GTM extensions inherently dictates along what lines will the data projections/representations be organized on the visualization plot. Loosely speaking, two structured data items will be mapped close to each other on the visualization plot, if they are likely to have been generated by the same underlying local noise model. This enables the user to have a control over the topographic organization via the choice and refinement of a noise model that reflects the user's beliefs and interests on data similarity. For example, instead of employing HMTMs as noise models, we could have formulated a different model with different modeling assumptions; while in HMTMs the position of children nodes of a node u is not significant, we could incorporate this information by making the transition/emission distribution of each child node depended on its position under the parent node u. As another example, an extension of GTM endowed with local noise models on tree-structured data, formulated with observable states, is presented in [16].

The second important advantage of the presented framework for model based visualization of structured data is the natural way in which goodness of fit and generalization can be quantified. We stress, that for recursive extensions of SOM, it is not clear how would one objectively assess the goodness of fit to a given data set. However, the ability to detect potential overfitting is crucial when dealing with nonlinear models. In our framework, we can calculate model likelihood on the training and/or hold-out validation data sets. Indeed, as shown in [16,20], such constrained mixture models with relatively wide basis functions on the latent space tend to be resilient to overfitting.

In the proposed framework it is possible to understand, quantify and visualize metric relationships among the structured data. The closeness of structured data items is induced by the local noise models used in the topographic map formulation. As mentioned above, two data items are close provided they are likely to have been generated from the same underlying local noise model. Again it is difficult to assess and/or understand metric relations among structured data items when recursive topographic map formulations are used.

Finally, the framework can be easily modified to perform structured data clustering instead of topographic visualization. As shown in [28], it is sufficient to change organization of the latent space. Briefly, for obtaining C clusters we need to use a discrete latent space $\mathcal{V}$ containing standard orthonormal basis of $\mathbb{R}^C$: $\mathcal{V} = \{x_1, x_2, ..., x_C\}$, where x_c, $c = 1, 2, ..., C$ is the C-dimensional vector of 0's except for coordinate c with value 1.

8   Conclusions

We have presented a principled probabilistic framework for topographic mapping of structured data. This work draws its inspiration from the GTM algorithm formulated for vectorial data. We formulated our models as mixtures of noise models on structured data (HMM/HMTM) that are parametrically constrained using a lower dimensional latent space $\mathcal{V}$. This induces a smooth low dimensional manifold $\mathcal{M}$ of noise models in the space of all noise models of the same form. Training is performed via the EM algorithm and can be visualized as folding and stretching of manifold $\mathcal{M}$ in response to the adjustment of parameters of the RBF mapping Γ in order to fit the data. Once the model is trained, each data point is projected to a location on the map corresponding to the mean of the posterior probability (responsibility) of latent space centers, given the data point. Alternatively, one may project to the latent center with the highest responsibility for the given data point (mode of the posterior). The projections on the map can be explained by inspecting plots of the parameters of the underlying noise models and reasoning about their generative behavior. Thus, two data items are placed close on the map if they are highly probable under similar local noise models. Furthermore, the calculation of magnification factors enhances our understanding of the maps by allowing us to appreciate the expansions/contractions on manifold $\mathcal{M}$ that occur due to the nonlinearity of the mapping Γ.

References

1. Kohonen, T.: The self-organizing map. Proceedings of the IEEE 78(9), 1464–1480 (1990)
2. Hammer, B., Micheli, A., Strickert, M., Sperduti, A.: A general framework for unsupervised processing of structured data. Neurocomputing 57, 3–35 (2004)
3. Svensén, M.: GTM: The Generative Topographic Mapping. Ph.D thesis, Aston University, UK (1998)
4. Kabán, A., Girolami, M.: A combined latent class and trait model for the analysis and visualization of discrete data. IEEE Transactions on Pattern Analysis and Machine Intelligence 23(8), 859–872 (2001)
5. Kleiberg, E., van de Wetering, H., van Wijk, J.J.: Botanical visualization of huge hierarchies. In: IEEE Symposium on Information Visualization, INFOVIS, pp. 87–94 (2001)
6. Chappell, G., Taylor, J.: The temporal kohonen map. Neural Networks 6, 441–445 (1993)
7. Koskela, T., Heikkonen, M.V.z.J., Kaski, K.: Recurrent SOM with local linear models in time series prediction. In: 6th European Symposium on Artificial Neural Networks, pp. 167–172 (1998)
8. Horio, K., Yamakawa, T.: Feedback self-organizing map and its application to spatio-temporal pattern classification. International Journal of Computational Intelligence and Applications 1(1), 1–18 (2001)
9. Voegtlin, T.: Recursive self-organizing maps. Neural Networks 15(8-9), 979–991 (2002)
10. Strickert, M., Hammer, B.: Merge SOM for temporal data. Neurocomputing 64, 39–71 (2005)
11. Hagenbuchner, M., Sperduti, A., Tsoi, A.C.: A self-organizing map for adaptive processing of structured data. IEEE Transactions on Neural Networks 14(3), 491–505 (2003)

12. Hagenbuchner, M., Sperduti, A., Tsoi, A.C.: Contextual processing of graphs using self-organizing maps. In: Proceedings of the European Symposium on Artificial Neural Networks (ESANN), pp. 399–404 (2005)
13. Heskes, T.: Energy functions for self-organizing maps. In: Oja, S., Kaski, E. (eds.) Kohonen Maps, pp. 303–315. Elsevier, Amsterdam (1999)
14. Bishop, C.M., Svensén, M., Williams, C.K.I.: GTM: The generative topographic mapping. Neural Computation 10(1), 215–234 (1998)
15. Rabiner, L.R.: A tutorial on hidden markov models and selected applications in speech recognition. Proceedings of the IEEE 77(2), 257–286 (1989)
16. Gianniotis, N., Tiňo, P.: Visualisation of tree-structured data through generative topographic mapping. IEEE Transactions on Neural Networks (2008) (in press)
17. Crouse, M., Nowak, R., Baraniuk, R.: Wavelet -Based Statistical Signal Processing Using Hidden Markov Models. IEEE Transactions on Signal Processing 46(4), 886–902 (1998)
18. Durand, J.B.: Gonçalvès, P.: Statistical inference for hidden Markov tree models and application to wavelet trees. Technical Report 4248, INRIA (2001)
19. Durand, J.B., Gonçalvès, P., Guedon, Y.: Computational methods for hidden markov tree models-an application to wavelet trees. IEEE Transactions on Signal Processing 52(9), 2552–2560 (2004)
20. Tiňo, P., Kaban, A., Sun, Y.: A generative probabilistic approach to visualizing sets of symbolic sequences. In: KDD 2004: Proceedings of the tenth ACM SIGKDD international conference on Knowledge discovery and data mining, pp. 701–706. ACM Press, New York (2004)
21. Bishop, C.M., Svensén, M., Williams, C.K.I.: Magnification factors for the gtm algorithm. In: Proceedings IEE Fifth International Conference on Artificial Neural Networks, pp. 64–69 (1997)
22. Cover, T.M., Thomas, J.A.: Elements of information theory. Wiley-Interscience, Hoboken (1991)
23. Kullback, S.: Information theory and statistics. Wiley, New York (1959)
24. Tiňo, P., Gianniotis, N.: Metric properties of structured data visualizations through generative probabilistic modeling. In: IJCAI 2007: 20th International Joint Conference on Artificial Intelligence, pp. 1083–1088. AAAI Press, Menlo Park (2007)
25. Do, M.N.: Fast approximation of Kullback-Leibler distance for dependence trees and hidden markov models. IEEE Signal Processing Letters 10(4), 115–118 (2003)
26. Merz, C., Murphy, P.: UCI repository of machine learning databases (1998)
27. Hagenbuchner, M., Tsoi, A.: The traffic policeman benchmark. In: Verleysen, M. (ed.) European Symposium on Artificial Neural Networks, April 1999, pp. 63–68. D-Facto (1999)
28. Kabán, A., Girolami, M.: A combined latent class and trait model for the analysis and visualization of discrete data. IEEE Transactions on Pattern Analysis and Machine Intelligence 23(8), 859–872 (2001)

Learning Highly Structured Manifolds: Harnessing the Power of SOMs

Erzsébet Merényi, Kadim Tasdemir, and Lili Zhang

Department of Electrical and Computer Engineering, Rice University, Houston, Texas, U.S.A.

Abstract. In this paper we elaborate on the challenges of learning manifolds that have many relevant clusters, and where the clusters can have widely varying statistics. We call such data manifolds *highly structured*. We describe approaches to structure identification through self-organized learning, in the context of such data. We present some of our recently developed methods to show that self-organizing neural maps contain a great deal of information that can be unleashed and put to use to achieve detailed and accurate learning of highly structured manifolds, and we also offer some comparisons with existing clustering methods on real data.

1 The Challenges of Learning Highly Structured Manifolds

Data collected today are often high-dimensional due to the vast number of attributes that are of interest for a given problem, and which advanced instrumentation and computerized systems are capable of acquiring and managing. Owing to the large number of attributes that are designed to provide sophisticated characterization of the problem, the data acquired are not only high-dimensional but also highly structured, i.e., the data have many clusters that are meaningful for the given application. Examples are hyperspectral imagery of planetary surfaces or biological tissues, DNA and protein microarrays, data bases for business operations and for security screening. These types of data created new demands for information extraction methods in regard to the detail that is expected to be identified. For example, hyperspectral imagery affords discrimination among many materials such as individual plant species, soil constituents, the paints of specific makes of cars, or a large variety of roof and building materials, creating a demand to extract as many as a hundred different clusters from a single remote sensing image of an urban scene. These clusters can be extremely variable in size, shape, density and other properties as we illustrate below. Another demand arising from such sophisticated data is to differentiate among clusters that have subtle differences, as the ability to do so can enable important discoveries. These examples highlight challenges for which many existing clustering and classification methods are not well prepared.

There has been much research on manifold learning motivated by the idea that the data samples, even if they are high-dimensional, can be represented by a low-dimensional submanifold. Representing the data in low-dimensional (2-d or 3-d)

M. Biehl et al.: (Eds.): Similarity-Based Clustering, LNAI 5400, pp. 138–168, 2009.

spaces can also help visualize the data structure to guide the user for capturing the clusters interactively. A classical technique for dimensionality reduction is principal component analysis (PCA) which works well when the data points lie on a linear submanifold. However, real data often lie on nonlinear spaces. To find the nonlinear subspaces (manifolds), many methods have been introduced (see [1,2] for recent reviews), among them a number of manifold learning algorithms: multidimensional scaling (MDS) [3], Isomap [4], locally linear embedding (LLE) [5], Hessian LLE (hLLE) [6] are some. These methods may successfully be applied to data sets that are characterized by only a few parameters (such as the angle of rotation in a number of similar video and image data sets) [7]. However, the same are often suboptimal for clustering applications as shown in various papers [8,9,10], since they are developed for reconstruction of one underlying submanifold rather than for identification of different groups in the data. In order to make manifold learning algorithms effective for classification, various works extend them with the help of the class statistics. [8] uses Fisher linear discriminant analysis (LDA) with Isomap for face recognition. Similarly, [11] uses LDA with manifold learning algorithms for face and character recognition. [10] modifies Isomap and LLE so that both local and global distances are considered for better visualization and classification. However, the performance of the modified Isomap or the modified LLE is not very promising due to the same reconstruction objective (the reconstruction of one underlying manifold) as for Isomap and LLE. In clustering applications, the aim is to learn the cluster structure — where the clusters may lie in different submanifolds — rather than to find one underlying submanifold for the data. Therefore representation of the separation between clusters is of great interest but not so much the precise topography of the underlying manifold. This makes adaptive vector quantization algorithms – which show the data topography on the prototype level and aim to faithfully represent the local similarities of the quantization prototypes – well suited for clustering [12,13,14,15,16].

Adaptive vector quantization algorithms are either inspired by nature as in the case of self-organizing maps (SOMs) [12], derived as stochastic gradient descent from a cost function as in Neural Gas [13] and its batch version [14], or derived through expectation-maximization [15,16]. Variants of SOM, Neural Gas and batch Neural Gas, which use a magnification factor in quantization, were also introduced and analyzed to control the areal representation of clusters (e.g., the enhancement of small clusters), in the learning process [17,18,19,20].

Among adaptive vector quantization methods we focus on SOMs. Our motivation is that not only can SOMs demonstratedly find optimal placement of prototypes in a high-dimensional manifold (and through that convey knowledge of the manifold structure) but the ordered prototypes also allow interesting and in-depth knowledge representation, regardless of the input dimensionality, which in turn helps resolve a large variety of clusters in great detail. After a brief background on SOM learning in Sect. 2, we discuss aspects of SOM learning as related to large high-dimensional manifolds with many clusters: quantification of the quality of learning (topology preservation) in Sect. 3.1; representation of the SOM knowledge, and extraction of clusters under these circumstances in Sect. 3.2, and we describe methods we developed in recent years. In Sect. 4 we present real data analyses, and offer conclusions in Sect. 5.

2 Learning Manifolds with Self-Organizing Maps

Self-Organizing Maps (SOMs) occupy a special place in manifold learning. They perform two acts simultaneously, during an iterative learning process. One is an adaptive vector quantization, which — assuming correct learning — spreads the quantization prototypes throughout the manifold such that they best represent the data distribution, within the constraints of the given SOM variant. (For example, the Conscience SOM produces an optimum placement of the prototypes in an information theoretical sense [21,18].) The other act is the organization (indexing) of the prototypes on a rigid low-dimensional grid, according to the similarities among the prototypes as measured by the metric of the data space. This duality makes SOMs unique among vector quantizers, and unique among manifold learning methods, because the density distribution — and therefore the structure — of a high-dimensional manifold can be mapped (and visualized) on a 1-, 2- or 3-dimensional grid without reducing the dimensionality of the data vectors. This, in principle, allows capture of clusters in high-dimensional space, which in turn facilitates identification of potentially complicated cluster structure that is often an attribute of high-dimensional data.

However, "the devil is in the details". The aim of this paper is to illuminate and quantify some of the details that are different for simple data and for complicated (highly structured) data, and to describe our contributions that alleviate certain limitations in existing SOM approaches (including the interpretation of the learned map) for highly structured manifolds.

For a comprehensive review of the SOM algorithm, see [12]. To briefly summarize, it is an unsupervised neural learning paradigm that maps a data manifold $M \subset \mathbb{R}^d$ to prototype (weight) vectors attached to neural units and indexed in a lower dimensional fixed lattice A of N neural units. The weight vector w_i of each neural unit i is adapted iteratively as originally defined by Kohonen [12]: Find the best matching unit i for a given data vector $v \in M$, such that

$$\|v - w_i\| \leq \|v - w_j\| \ \forall j \in A \tag{1}$$

and update the weight vector w_i and its neighbors according to

$$w_j(t+1) = w_j(t) + \alpha(t)h_{i,j}(t)(v - w_j(t)) \tag{2}$$

where t is time, $\alpha(t)$ is a learning parameter and $h_{i,j}(t)$ is a neighborhood function, often defined by a Gaussian kernel around the best matching unit w_i. Through repeated application of the above steps, the weight vectors become the vector quantization prototypes of the input space M.

This is an enigmatic paradigm: it has been studied extensively (e.g., [12,22,23,24,25]), yet theoretical results are lacking for proof of convergence and ordering for the general case. In principle, the SOM is a topology preserving mapping, *i.e.*, the prototypes which are neighbors in A should also be neighbors (centroids of neighboring Voronoi polyhedra as defined in [26]) in M, and vice versa. When centroids of neighboring Voronoi polyhedra in M are not neighbored in the SOM lattice we speak of *forward topology violation* with a folding length k where k is the lattice distance between the prototypes in question. When two SOM neighbors are not centroids of adjacent Voronoi polyhedra in M a *backward topology violation* occurs [27]. Both conditions can be (and are usually) present at various stages of SOM learning, to various extent depending on the characteristics of the data and the SOM lattice. This is so even if the learning "goes well" (no twists develop in the map), and topology violations may not vanish with any amount of learning. One is therefore motivated to construct empirical and heuristic measures to quantify the quality of learning: the *degree* of faithfulness and maturity of the mapping, as it is meaningful for a given application. Measures of topology preservation, such as the Topographic Product, Topographic Error, and Topographic Function, have been proposed [28,29,27], which define perfect mapping in exact numerical terms. The Topographic Function [27] also shows topology violations as a function of the folding length, which gives a sense of how global or local the violations are on average. What these existing measures do not provide, however, is a clear sense of what the numbers mean for less than perfect learning: how far the score of an already usefully organized state of the SOM could be from perfect (zero for the Topographic Product and the Topographic Error); whether a numerical value closer to the perfect score necessarily means better organization; or which of two Topographic Functions express better organization.

The quality of learning can be viewed in relation to the goal of the learning. Two levels are easily distinguished: a) the learning of the topography (the density distribution); and b) the learning of the cluster structure, of a manifold for which the acceptable degree of topographic faithfulness can be quite different. Learning cluster structure does not require a very precise learning of the topography. Certain level of local topology violations in the SOM is tolerable and will not hinder the accurate extraction of clusters. To illustrate the point, consider the SOM learned with a relatively simple data set, in Fig. 1. The data set contains 8 classes, in a 6-dimensional feature space. The classes are apportioned such that four of them comprise 1024 data vectors each, two of them have 2048, and two have 4096 data vectors. Gaussian noise, about 10% on average, was added to the data vetors to create variations within the classes. The clusters detected by the SOM are delineated by the white "fences" we call the mU-matrix, and by the empty (black) prototypes, as explained in the figure caption. (We will elaborate on this particular knowledge representation more in Sect. 3.2.) Comparison with the known class labels (colors) superimposed in Fig. 1, right, makes it obvious that the SOM already learned the cluster structure perfectly. At the same time, one can examine the topology violations and conclude that the topography is far from having been learned perfectly. To show this we connected with black lines those prototypes which are not neighbors in the SOM grid but have adjacent Voronoi cells in data space. These lines express the forward topology violations,

 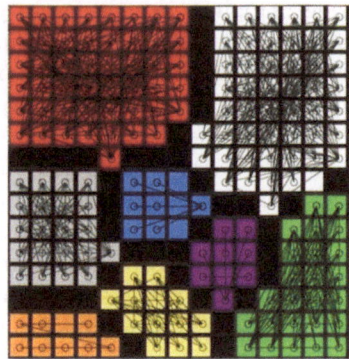

Fig. 1. Left: 15 x 15 SOM of an 8-class 6-dimensional synthetic data set. The cluster boundaries emerge through the mU-matrix, the visualization of the distances of adjacent prototypes as fences between the grid cells in all eight directions. White is high fence (large dissimilarity), black is low fence (great similarity). Each grid cell is shaded by an intensity of red proportional to the number of data points mapped to the prototype in that grid cell. Black grid cells indicate empty receptive fields of the corresponding prototypes. The fences clearly outline 8 clusters. **Right:** The SOM of the 8-class synthetic data with known class labels (colors) superimposed. The clustering learned by the SOM is perfect in spite that many topology violations still exist at this stage of the learning, even at folding length 8 (i.e., nearly half the width of the SOM lattice). The (567) existing violations, shown by black lines in the cluster map, all occur locally within clusters.

and we call them *violating connections*. All violating connections are residing locally inside clusters, without confusing the cluster boundaries. Apparently, a coarser organization has taken place, separating the clusters, and finer ordering of the protoypes would continue within the already established clusters. One might conclude that, for the purpose of cluster capture, this level of topology violation is tolerable and inconsequential [30]. We elaborate on aspects of measuring topology violations, and present new measures, in Sect. 3.1 and 3.2.

3   Learning the Clusters in Highly Structured Manifolds

As stated in Sect. 1 high-dimensional data are often complicated, highly structured, as a result of the application task for which the data were collected. Complicated means the presence of many clusters which may not be linearly separable, and which can be widely varying in various aspects of their statistics, such as size, shape (non-symmetric, irregular), density (some are very sparse, others are dense in feature space), and their proximities. Fig. 2 gives an illustration of these conditions.

To give a real example of a situation similar to that in Fig. 2 we show statistics of a remote sensing spectral image of Ocean City, Maryland [31]. This data set is described in detail in Sect. 4.1. In previous analyses more than twenty clusters

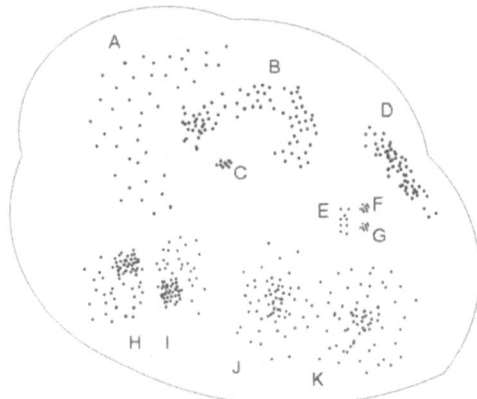

Fig. 2. Illustration of complicated cluster structure. The eleven clusters depicted here are in different proximity relations, and have different statistics. Their shapes vary from spherical (F, G, K) through ellipsoidal (C, D) and ellipsoid-like (H, I, J) to rectangular (E) and irregular (A, B). Some (B, C, D, F and G) are denser than others (A, E, J and K), and some (H and I) have inhomogeneous data distribution. Several (C, F, G) are very small, and two (J and K) are overlapping. Such variations produce a challenging structure that may not be effectively dealt with by methods which, by design, best handle clusters of given characteristics (such as spherical clusters by K-means).

were identified in this image and associated with meaningful physical entities such as roof materials, vegetation, roads, etc. [18]. A selected subset of those clusters is listed in Table 1, for the eight spectral bands (feature space dimensions) which were used in the earlier clustering. First, the number of data points varies extremely, from a few hundred to nearly 50,000, across clusters. Secondly, the standard deviation for each of the clusters varies across the spectral bands, and varies differently for each cluster (anywhere from 2- to 5-fold), indicating all kinds of shapes except hyperspherical. To compare the volumes of the clusters, we assume that clusters are hyperrectangles with a size proportional to the standard deviation in the corresponding dimensions (bands). While this does not give the correct volume of the clusters it still provides an insight to the magnitude of the variation. By this rough comparison, the largest cluster (L), is about 2000 times larger than the smallest cluster (G). We also compute the density of the clusters by dividing the number of data points by the calculated volume. The densest cluster (G) is about 3500 times more dense than the sparsest cluster (L). For example, G has about 28,000 data points, a volume of 4×10^{15} and a density of 7463×10^{-15} whereas I, which has a similar number of data points (about 25,000), has a volume of 211×10^{15} and a density of 117×10^{-15} ($1/70$ of the density of G). In Sect. 3.2 we will also show envelopes of these clusters (i.e., plots of extreme values in each feature dimension), from which it can be seen that many of the clusters are overlapping. The widely varying numbers of data points, volumes and densities, as well as overlaps of the clusters make this Ocean City data set complicated, highly structured.

Table 1. Statistics indicating complicated cluster structure in a remote sensing spectral image of Ocean City, Maryland. From over twenty identified clusters, corresponding to real physical entities in this image, nine selected clusters are listed here.

Cluster Labels	No. data points	standard deviation								volume (10^{15})	density (10^{-15})
		band 1	band 2	band 3	band 4	band 5	band 6	band 7	band 8		
A	8089	145	88	171	101	136	102	175	181	96	84
B	5496	186	141	118	90	122	92	145	146	65	84
E	14684	186	109	76	102	126	81	179	229	65	225
G	28468	122	73	84	65	85	60	107	145	4	7463
H	480	176	109	118	112	115	81	156	214	78	6
I	24719	271	107	174	189	157	86	119	138	211	117
L	13592	339	268	207	207	233	148	238	269	8589	2
O	20082	341	221	157	159	213	148	283	275	4610	4
R	48307	179	119	163	106	107	109	235	269	271	178
V	998	177	229	112	79	92	81	180	217	106	9
a	239	354	286	205	175	168	144	201	177	3132	0.1

Extraction of such complicated clusters, and rather precisely, is important in many of today's real problems. For example (as we show in Sect. 4.2) some of the smallest clusters in this image represent unique roofing materials, and many of the clusters with considerable overlap in their signatures map distinct man-made materials. All of these are important to detect and map accurately in an urban development survey (for which this kind of data may be acquired). We want to point out that this Ocean City image is not as complicated as some of the hyperspectral images we have been working with. One such image will also be analyzed in Sect. 4.3. In the rest of this paper we discuss our contributions to the learning and extraction of clusters from highly structured manifolds.

3.1 Measuring the Correctness of SOM Learning for Complicated High-Dimensional Manifolds

A prerequisite of faithful cluster identification from an SOM is an appropriate representation of the topology of the data space M by the ordering of the prototypes in the SOM lattice A. Ideally, the SOM should be free of topology violations, at least in the forward direction since the "twists" caused by forward violations can lead to incorrect clustering. Backward topology violations are not detrimental for cluster extraction because they manifest in disconnects (strong dissimilarities — high fences — and/or prototypes with empty receptive fields, as in Fig. 1) in the SOM, which helps locate clusters. Ideal topology preservation usually does not occur for real data. For noisy, complicated manifolds topology violations are common at all stages of the learning. Adding to the difficulty is the fact that one does not know when the learning is mature enough, or how much further a seemingly static map may still improve. The key is to quantify the extent of various violations and be able to separate important ones from the inconsequential.

The Topographic Product [28] (TP) was the first measure that quantified the quality of topology preservation. A prototype based measure, the TP is computationally economical, but it penalizes (falsely detected) violations caused by nonlinearities in the manifold, due to improper interpretation of neighborhood by Euclidean metric. This drawback is remedied in the Topographic Function [27] (TF), where the Euclidean metric is replaced by the graph metric of the induced Delaunay triangulation. Assuming a high enough density of prototypes in the manifold, the induced Delaunay graph can be constructed, after [26], by finding the best matching unit (BMU) and the second best matching unit (second BMU) for each data vector, and expressing these "connections" in a binary adjacency matrix of the SOM prototypes. Two prototypes that are a pair of BMU and second BMU for any data vector are adjacent or *connected* in the induced Delaunay graph. (Equivalently, they are centroids of adjacent Voronoi cells). The lattice (maximum norm) distance of two prototypes in the SOM is their *connection length* or *folding length* [27]. The TF not only uses a better distance metric, but also shows the scope of forward violations, by computing the average number of connections that exist at folding lengths larger than k:

$$TF(k) = \frac{1}{N} \sum_{i \in A} \{ \# \text{ of connections of unit } i \text{ with length} > k \}, \qquad (3)$$

where i is the index of the neural unit in A, N is the total number of units. A large k indicates a global (long range, more serious) violation while a small k corresponds to a local disorder. Eq. (3) is also applicable to backward violations, where k is negative and represents the induced Delaunay graph distance, in the data space, between prototypes that are adjacent in the SOM. There are also measures that only utilize information on the data distribution. For example, [32] introduced a cumulative histogram to express the stability of the neighborhood relations in an SOM. It captures a statistical view of the neighborhood status of the system and compares it with an unordered map. The more dissimilar they are, the more reliable the mapping. Another measure, the Topographic Error [29], expresses the extent of topology violation as a percentage of data points that contribute to violating connections. Extending these previous works, we enriched and further resolved the TF in the Weighted Differential Topographic Function (WDTF) [30].

$$WDTF(k) = \frac{1}{D} \{ \# \text{ of data vectors inducing connections of length} = k \} \quad (4)$$

where D is the total number of data samples. The WDTF is a differential view of the violations at different folding lengths, in contrast to the integral view of the TF. It also adds new information by using the number of data samples that induce a given connection, as an *importance weighting*. By this weighting, the WDTF distinguishes the severity of violating connections: a long range but weak violating connection caused by a few noisy data vectors may be unimportant and safely ignored in the overall assessment of topological health, while a heavy violating connection warrants attention as a potential twist in the map.

Monitoring Violating Connections: TopoView and WDTF. Besides the topographic measure WDTF discussed above, we present another useful, interactive tool, TopoView, which allows to show violating connections on the SOM. This is more general than displaying SOM neighbors connected in data space, which is limited to 2- or 3-dimensional data. Different subsets of connections can also be selected by thresholding the connection strength and/or the connection length, which helps filter out noise, outliers and unimportant (weak) violations, and thereby more clearly see the relevant characteristics of the topology. For complex and high-dimensional data, it is especially effective to use both tools together. The WDTF provides a summary of the severity of violations at each folding length while TopoView provides localization of the violations in the SOM, for selected severity levels.

We illustrate the use of TopoView and the WDTF on a synthetic 4-class Gaussian data set, in Fig. 3. The data set was generated by using four Gaussian distributions with mean=0, standard deviation=1, at four centers in 2-dimensional space. At 1K (1000) steps (Fig. 3, top row), the SOM appears twisted in the data space, especially in the upper right cluster, where a chain of SOM protoypes is arranged in the shape of a horseshoe. TopoView reflects this twisting by a set of connections along the right side of the SOM. From the WDTF, which shows violations up to length 6, we can see what is known from the connection statistics: the end units of this chain, and also some of the non-neighbored units in between, must be connected. However, the long range violations are relatively weak. The rest of the violating connections are more local, but stronger, with a connection length of 2 or 3. As the SOM evolves, the set of long range connections in the SOM disappear at 3K steps (middle row), which means the "horseshoe" took up a shape that better approximates the spherical cluster. Finally, TopoView shows the SOM free of violating connections at 100K steps (bottom row): the prototypes are well placed in the data space, and the WDTF vanishes.

We give a real demonstration of the use of these measures through a hyperspectral urban image, which represents complicated, highly structured data. This image, which we will call "RIT image", was synthetically generated, therefore it has ground truth for every pixel, allowing objective evaluation of analysis results on the 1-pixel scale. The image pixels are 210-dimensional feature vectors (reflectance spectra), which are the data vectors input to the SOM. The scene contains over 70 different material classes. Fig. 4 and 5 give an illustration, and Sect. 4.1 a detailed description, of this data set.

To monitor topology preservation we compare two snapshots taken during the learning of the RIT image, at 500K and 3M (3000000) steps, respectively. Fig. 6 shows that the quality of the topology preservation improves from 500K to 3M steps. The number of short-range violations considerably decreases, and a decrease is generally showing at larger folding lengths, with some exceptions (such as at $k = 8$ and $k = 13$). From the TopoView representation in Fig. 7 one can follow which violations disappear between the two snapshots. Since TopoView shows the individual violating connections the thick cloud of connections can be

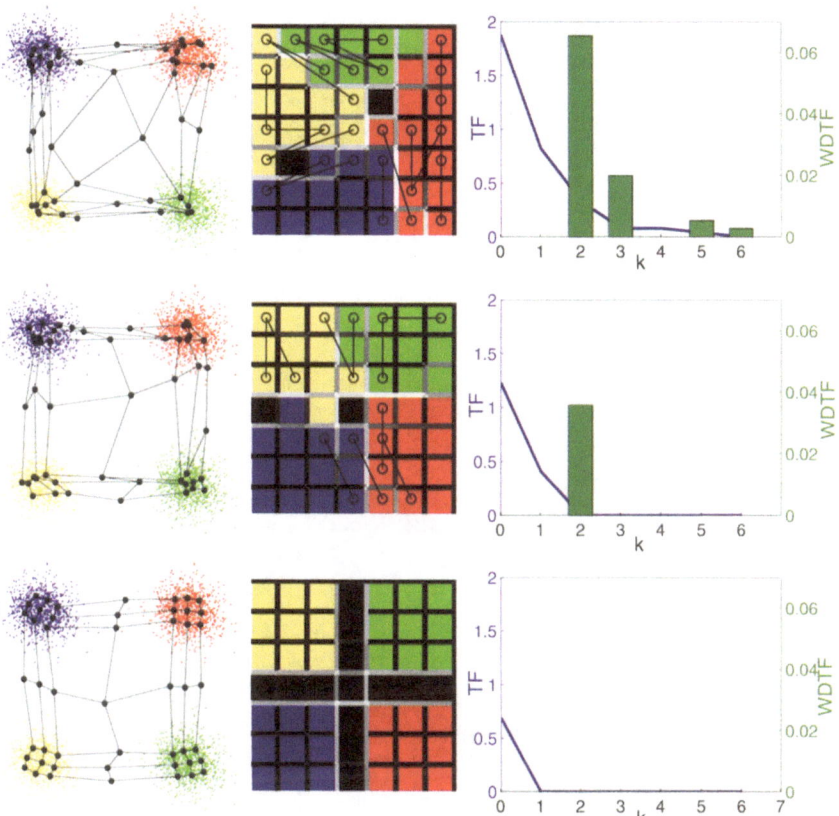

Fig. 3. The evolution of the SOM as it learns the synthetic 2-dimensional 4-class Gaussian data. Three snapshots are shown at 1K, 3K and 100K steps, from top to bottom. **Left**: The SOM prototypes (black dots) in the data space space, with SOM neighbors connected. Data samples are color coded according to their class memberships. **Center**: All violating connections shown as black lines, over the SOM, with the class labels (colors) and the mU-matrix also superimposed. **Right**: The TFs (blue lines) and the WDTFs (green bars).

obscuring. However, even with all connections shown (in the top row of Fig. 7), one can see that, for example, the lower left corner of the SOM became completely violation free at 3M steps. To give an "importance-weighted" view of the same we can apply tresholding by connection strength (the weighting used by WDTF). This eliminates unimportant connections and clears the view for analysis of those violations that may significantly contribute to cluster confusion. Two examples for possible thresholdings are given in Fig. 7. In both cases a decrease in confusion (relative to the mU-matrix fences) can be seen. Thresholding by connection length can make the cut between global and local range violations. This threshold depends on the data statistics and is automatically computed, based on the following argument, from [34]: if a prototype w_i has m Voronoi neighbors in data space

Fig. 4. Top: A color composite of 3 selected spectral bands of the 400 x 400 pixel, 210-band synthetic hyperspectral RIT image. **Bottom:** Partial cluster map, extracted from the self-organizing map in Fig. 5, using the modified U-matrix visualization [33]. It shows 39 cover types with unique colors keyed in the color wedge. Some additional unique materials (such as roofs of houses, showing in black) have no labels assigned for reasons of color limitations. Besides the obvious vegetation (trees and grasses, clusters I, J, K, Z, g, Y), the approximately 70 different surface cover types in this image include mixed dirt/grass (T), a large number of roof materials (A, B, C, F, M, N, Q, R, U, X, a, b, d, h, i, j, k, l, m), pavings (V, roads and parking lots), tennis courts (O, P), several types of car paints (W, c, M, f, e), and glass (windshield of cars and roof, E).

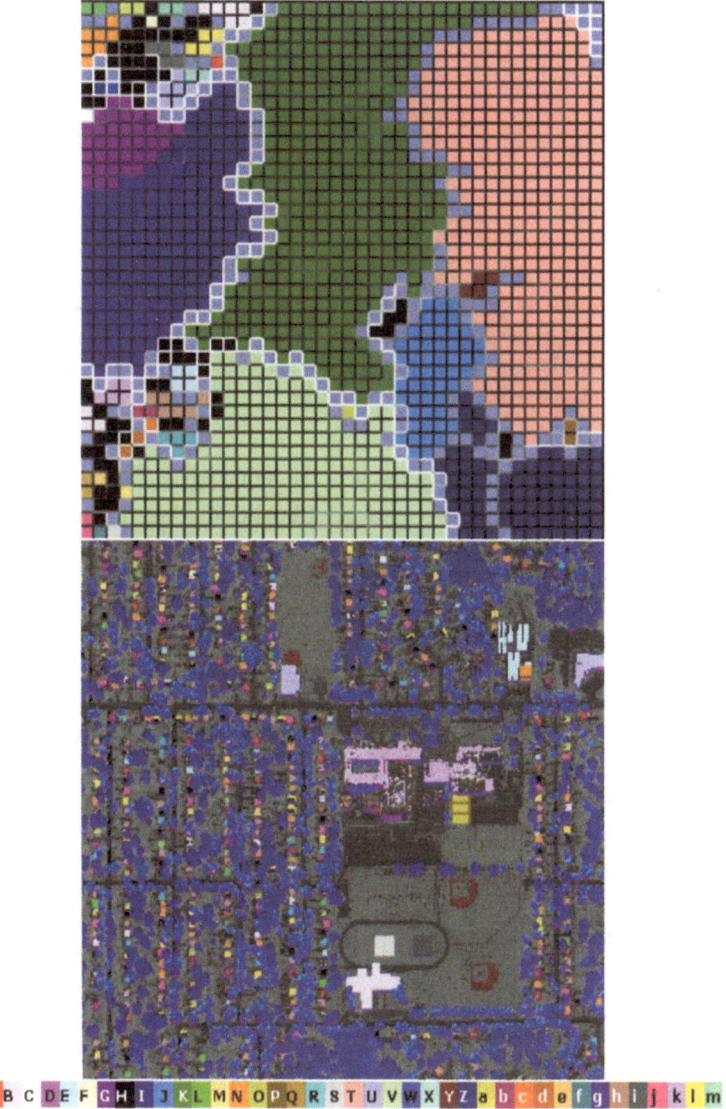

Fig. 5. Top: The SOM with discovered clusters color-coded and the mU-matrix superimposed. Interpretation of these clusters is given in Fig. 4. Medium grey cells (the color of the background, "bg"), appearing mostly along cluster boundaries are SOM prototypes with no data points mapped to them. Some prototypes — shown as black cells, which have data mapped to them — were left unclustered, because of color limitation. **Bottom:** The same cluster map as in Fig. 4, bottom, with the large background clusters (grass, paved roads and lots) removed to provide better contrast for the many different roof materials, and other small unique spectral clusters such as tennis courts. About twenty of these clusters are roof types. Spectral plots showing excellent match of the spectral characteristics of the extracted clusters with true classes are in [33].

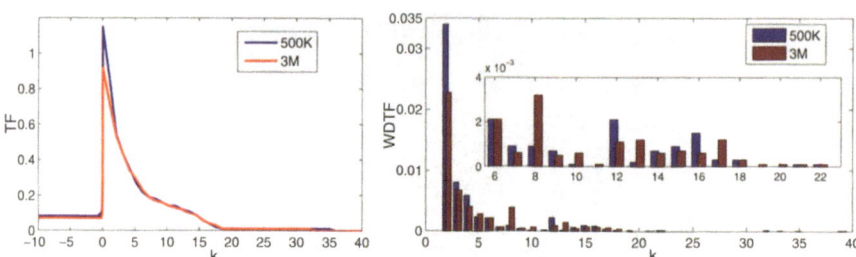

Fig. 6. Comparison of TFs (**left**) and WDTFs (**right**) after 500K and 3M SOM learning steps on the RIT hyperspectral image. While the integral measure TF shows a general decrease of violations at shorter folding lengths, the WDTF also indicates the fluctuations of violations across folding lengths.

then a topology preserving arrangement in the SOM lattice for these m neighbors is a placement into the "tightest" SOM neighborhood. This means that the 8 closest Voronoi neighbors should occupy the 8 immediate lattice neighbors of w_i in a square SOM neighborhood, Voronoi neighbors 9 – 24 wrap around this first tier of immediate SOM neighborhood, and so on. The radius of the SOM neighborhood that accomodates all Voronoi neighbors in this tightest fashion yields the folding length k within which the violations can be considered local. In the case of the RIT data, the maximum number of connected neighbors is 21 at 500K steps and 19 at 3M steps. These can fit into a 2-tier (8+16) square SOM neighborhood. Therefore, global violations are those with $k \geq 3$, shown in Fig. 7, bottom row. TopoView can also show inter-cluster or intra-cluster connections separately if a clustering is provided, and thus aid in the verification of clustering. This will be shown in Sect. 4.3.

3.2   Cluster Extraction

Cluster extraction from SOMs is accomplished through the clustering of the learned prototypes. Various approaches can be used, either based solely on the similarities of the prototypes or by taking into account both the prototype similarities and their neighborhood relations in the SOM grid. The latter is typically done interactively from visualization of the SOM and is generally more successful in extracting relevant detail than automated clustering of the prototypes with currently available methods.

SOM knowledge representations — what information is quantified and how — are key to the quality of cluster capture from visualizations. The widely used U-matrix [35] displays the weight distances of the SOM neighbor prototypes, averaged over the neighbors and coloring the grid cell of the current prototype to a grey level proportional to this average distance. The U-matrix and its variants (e.g., [36], [37], [38]) are most effective when relatively large SOM grid accomodates small data sets with a low number of clusters because the averaging can obscure very small clusters and sharp boundaries in a tightly packed

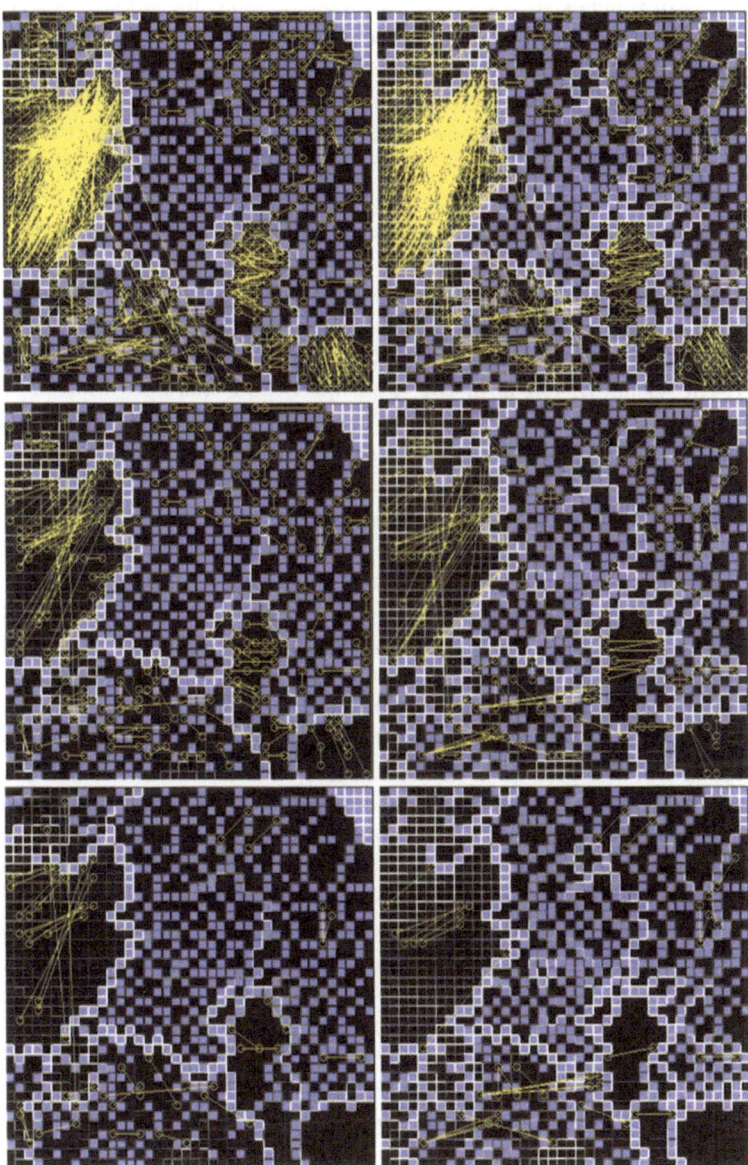

Fig. 7. TopoView visualization of violating connections (yellow lines) with different thresholdings, on the SOM of the RIT data at 500K steps (left column) and at 3M steps (right column). In the underlying SOM, the mU-matrix is superimposed (white fences). Medium grey cells indicate empty prototypes. **Top:** all violating connections are drawn. **Center:** violating connections with strength greater than the mean strength of all violating connections. **Bottom:** global violating connections ($k \geq 3$) with connection strength greater than the mean strength of the fourth strongest connections of all prototypes. This choice of threshold, proposed in [34], is described in Sect. 3.1.

SOM grid. Approaches such as [39] and gravitational methods (e.g., Adaptive Coordinates [37]) visualize distances between the weights in innovative ways that greatly help manual cluster extraction. Automated color assignments also help qualitative exploration of the cluster structure [40], [41], [42]. We point the reader to [37], [43], [34] for review. Visualization of the size of the proto-type receptive fields (e.g., [38], [44]) is among the earliest tools. Visualization of samples that are adjacent in data space but map to different SOM proto-types [45] is a richer representation than the previous ones since it makes use of the data topology. However, in case of a large number of data points, adjacent samples mapped to different prototypes are only the ones at the boundaries of the Voronoi polyhedra, thus this visualization still leaves a lot of the topological information untapped. More of the data topology is utilized by [46] and [47], however, the visualization is in the data space and therefore limited to up to 3 dimensions.

We added to this arsenal the *modified U-matrix* (mU-matrix), and the *connectivity matrix*, CONN, and their visualizations. They are especially useful for large, highly structured data sets mapped to not very large SOMs. (The size of the SOM is a sensitive issue with high-dimensional data as the computational burden increases non-linearly with input dimension. One wants to use a large enough SOM to allow resolution of the many clusters potentially present in the data set, but not exceedingly larger than that.) The mU-matrix is a higher res-olution version of the U-matrix in that it displays the weight distances to all neighbors separately, on the border of the grid cells including the diagonals, as shown in Fig. 1. Combined with the representation of the receptive field sizes (red intesities of the grid cells in Fig. 1, left), it conveys the same knowledge as in [39]. While the sense of distance in the mU-matrix is not as expressive as the "carved away" grid cells in [39] the mU-matrix leaves room for additional information to be layered. Examples of that are [48] and [49] where known labels of individual data objects were displayed in the grid cells thereby showing, in addition to the density, the distribution of the known classes within receptive fields. The mU-matrix is advantageous for the detection of very small clusters, as for example, in Fig. 5, where many small clusters are represented by just a few (even single) protoypes in the upper and lower left corners.

The CONN knowledge representation was first proposed in [50], developed for visualization in [51], and is presented in detail in [34]. It is an extension of the induced Delaunay triangulation, by assignment of weights to the edges of the graph. An edge connecting two prototypes is weighted by the number of data samples for which these prototypes are a BMU and second BMU pair. This weighting is motivated by the unisotropic distribution of the data points within the Voroni cells, as explained in Fig. 8 on the "Clown" data created by [52].

The edges of the weighted Delaunay graph can be described by

$$CONN(i,j) = |RF_{ij}| + |RF_{ji}| \tag{5}$$

where RF_{ij} is that section of the receptive field of w_i where w_j is the second BMU, and $|RF_{ij}|$ is the number of data vectors in RF_{ij}. Obviously, $|RF_i| =$

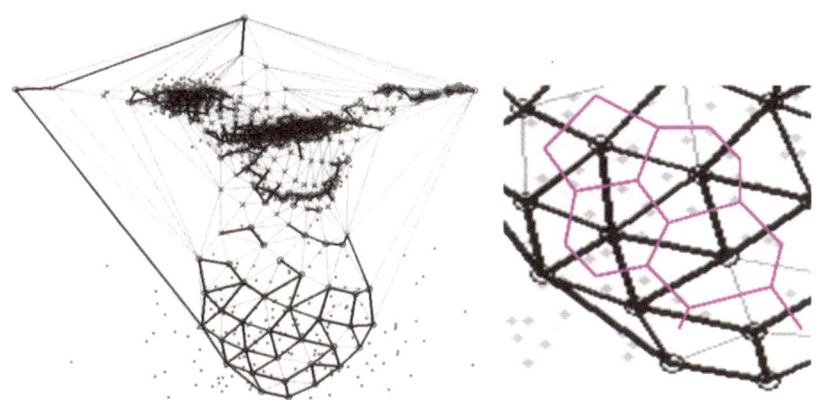

Fig. 8. Left: Delaunay triangulation (thin lines) and induced Delaunay triangulation (thick lines) for the 2-d "Clown" data set, using the SOM prototypes created by [52]. The "Clown" is indicated by the annotations in Fig. 9. We omit annotations here so as not to obscure details. The small dots represent data points. Prototypes with non-empty receptive fields are denoted by circles, prototypes with empty receptive fields are shown by 'x'. The induced Delaunay triangulation exposes discontinuities in the data manifold, for example, the separations between the eyes, the nose and the mouth, while the Delaunay triangulation does not highlight them. **Right:** Magnified detail from the lower left part of the "Clown". Data points in the Voronoi cells, superimposed in pink, exhibit an unisotropic distribution, indicating variable local data densities in the directions of the Voronoi neighbors.

$\sum_{j=1}^{N} |RF_{ij}|$ because $RF_i = \cup_{j=1}^{N} RF_{ij}$. CONN thus shows how the data is distributed within the receptive fields with respect to neighbor prototypes. This provides a finer density distribution than other existing density representations which show the distribution only on the receptive field level. $CONN(i,j)$, the connectivity strength, defines a similarity measure of two prototypes w_i and w_j.

Visualization of this weighted graph, CONNvis, is produced by connecting prototypes with edges whose widths are proportional to their weights. The line width gives a sense of the *global importance* of each connection because it allows to see its strength in comparison to all other connections. A *ranking* of the connectivity strengths of w_i reveals the most-to-least dense regions local to w_i in data space. This is coded by line colors, red, blue, green, yellow and dark to light gray levels, in descending order. The ranking gives the relative contribution of each neighbor independent of the size of w_i's receptive field, thus the line colors express the *local importance* of the connections. The line width and the line color together produce a view of the connectedness of the manifold, on both global and local scales. This is shown in Fig. 9 for the "Clown" data. We use this 2-dimensional data set because it has an interesting cluster structure, and because we are able to show the information represented by CONN both in data space and on the SOM, thus we can illustrate how CONNvis shows data structure on the SOM regardless of the data dimension. Compared to Fig. 8, left, all connections remain, but now the connection strengths emphasize strongly and

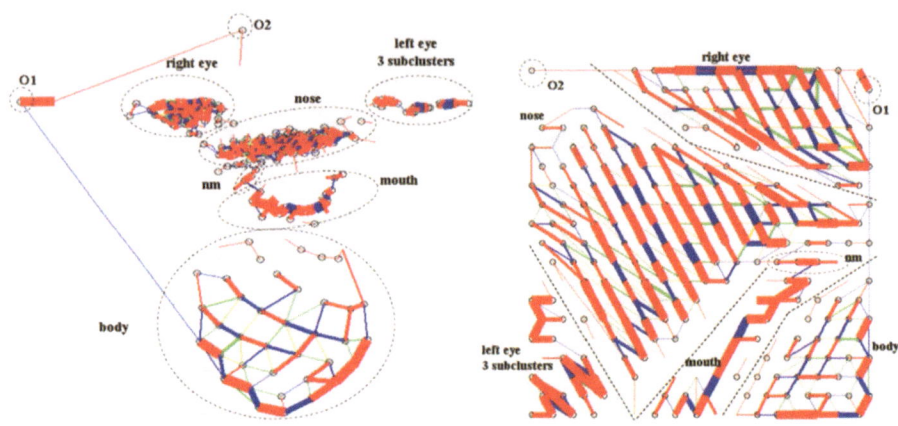

Fig. 9. Left: The connectivity matrix CONN (weighted Delaunay triangulation) shown on the 2-d "Clown" data set, using the SOM prototypes created by [52]. Parts of the "Clown" are explained by the annotations. O1 and O2 are outliers. The lack of a circle symbol indicates an empty prototype. Line widths are proportional to the weight of the edge, $CONN(i,j)$, expressing its global importance among all connections. Colors show the ranking, the local importance of the connections to the Voronoi neighbors. This is redundant with the line widths but because we bin the line widths to help the human eye distinguish grades of connection strengths, color coding the ranking restores some of the information lost through the binning. The ranking is not symmetric, *i.e.,* if the rank of w_j for w_i is r, and the rank of w_i for w_j is s, r is not necessarily equal to s. The connections are drawn in the order of lowest to highest rank so a higher-ranking connection will overlay a lower-ranking one. Details of subcluster structures are visibly improved compared to Fig. 8. **Right:** The same CONN matrix draped over the SOM.

poorly connected (high and low density) regions. Clusters not obvious in Fig. 8 and not visible in the U-matrix of the Clown in [52] such as the three subclusters in the left eye, clearly emerge here.

One significant merit of CONNvis is that it shows forward topology violations in a weighted manner, giving a strong visual impression of the densest textures in the data. CONNvis is somewhat limited in resolving many connections because the weighting (line width) can obscure finer lines in a busy CONNvis. Complementarily, TopoView can show many connections simultaneously, in selected ranges of the connection strength. Alternative use of these two visualizations, which render the same knowledge, can be quite powerful. Both CONNvis and TopoView also show backward violations through unconnected SOM neighbors. These indicate discontinuities in the manifold and thus immediately outline major partitions in the data. The mU-matrix has capacity to indicate backward topology violations through corridors of prototypes with empty receptive fields, such as in Fig. 1, and through high fences but not as clearly as CONNvis or TopoView, and it cannot show forward violations.

Cluster Extraction with CONNvis. Interactive clustering with the help of CONNvis can be done by evaluation and pruning of the connections. Unconnected

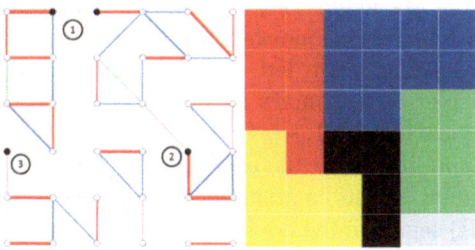

Fig. 10. Left: Illustration of interactive clustering with CONNvis. Strongly connected groups of prototypes (coarse clusters) visually emerge through the lack of connections or weak connections across some prototypes (annotated by black dots), easily recognized by a human analyst. The connections of these straddling prototypes present one of three situations for which the following rules are applied: 1: If a prototype has different number of connections to each coarse cluster, cut the smaller set of connections; 2: If a prototype has the same number of connections to each cluster but with different connectivity strengths, cut the weaker set of connections; 3: If a prototype has the same number of connections to each cluster, with the same strengths but different rankings, cut the lower ranking ones. **Right:** The identified clusters.

or weakly connected prototypes in the unpruned CONN often already outline "coarse" clusters, i.e., densely connected areas in the SOM which have considerably fewer connections to one another. The weakly connected prototypes at the boundaries of the coarse clusters are easily recognized by the human operator (black dots in Fig. 10). The corresponding weak connections are then evaluated as described in Fig. 10, to find and severe the "weakest link" for proper cluster separation.

For complicated cases, where the number of data points is huge, and the data are noisy, prototypes can have a large number of connections (neighbors in data space), and also a relatively large number of connections across coarse clusters. This creates a busy CONNvis, and requires considerable work when the number of clusters in the data is large, but the procedure is exactly the same as in simple cases. It has been used to produce some of the results shown here, and elsewhere [33]. At this time of writing we are collecting experiences from interactive clustering with CONNvis, which we expect to turn into an automatic procedure ultimately.

4   Case Studies with Higly Structured Real Data Sets

4.1   Real Data Sets

The Ocean City Multispectral Image. was obtained by a Daedalus AADS-1260 multispectral scanner. The image comprises 512×512 pixels with an average spatial resolution of 1.5 m/pixel [31]. Each pixel is an 8-dimensional feature vector (spectrum) of measured radiance values at the set of wavelengths in the 0.38–$1.1\mu m$ and 11–$14 \mu m$ windows that remained after preprocessing [53].

Ocean City, along the Maryland coast, consists of rows of closely spaced buildings separated by parallel roads and water canals. The spatial layout of different surface units is shown in Fig. 11, left, through an earlier supervised class map [53], which we consider a benchmark since it was verified through aerial photographs and field knowledge [31,54]. Ocean (blue, I) surrounds the city, ending in small bays (medium blue, J, at the top center and bottom center of the scene) which are surrounded by coastal marshlands (brown, P; ocher, Q). Shallow water canals (turquoise, R) separate the double rows of houses, trending in roughly N-S direction in the left of the scene and E-W direction in the right. Many houses have private docks for boats (flesh-colored pink, T) and as a consequence, dirty water at such locations (black, H). Paved roads (magenta, G) with reflective paint in the center (light blue, E) and houses with various roof materials (A, B, C, D, E, V) show as different classes. Typical vegetation types around buildings are healthy lawn, trees and bushes (K, L), yellowish lawn (split-pea green, O) and dry grass (orange, N).

The RIT Hyperspectral Image. briefly introduced in Sect. 3.1 and in Fig. 4 and 5, was synthetically generated through rigorous radiative transfer modelling called the DIRSIG procedure at the Rochester Institute of Technology [55,56]

Fig. 11. Left: Supervised classification of the Ocean City image, mapping 24 known cover types. Red, white and blue ovals show unclassified shapes of buildings and a circle at the end of a road (the color of the background, bg). **Right:** Clusters identified interactively from CONNvis visualization of a SOM of the Ocean City image. The agreement between the cluster map and supervised class map is very good. The unclassified gray spots (in red, white and blue ovals on the left) are now filled exactly, and with colors (a, c, j) different from the 24 colors of the supervised classes. These new clusters only occur at the locations shown, indicating the discovery of rare roof types and road materials, which were not used for the training of the supervised classifier. The spectral signatures of the newly discovered clusters, as well as those of two new subclusters (e, m) are distinct from the rest.

(hence the name RIT). Owing to its simulated nature, this image has ground truth for every pixel, allowing objective evaluation of analysis results on the 1-pixel scale. The realism of the RIT image is quite amazing, in both spectral and spatial respect. Its characteristics are close to that of an AVIRIS image. (AVIRIS is the Airborne Visible Near-Infrared Imaging Spectrometer of NASA/JPL [57,58], to date the most established, extremely well calibrated and understood airborne hyperspectral imager.) The scene comprises 400 x 400 pixels in 210 image bands in the 0.38 to 2.4 μm visible-near-infrared spectral window. The spatial resolution is 2 m/pixel. Realistic noise and illumination geometry is part of the simulation. The visual appearance of a natural color composite made from three selected image bands in the red, green, and blue wavelength regions, Fig. 4, top, is virtually indistinguishable from an image of a real scene. It contains over 70 different classes of surface materials, widely varying in their statistical properties in 210-dimensional data space. The materials include vegetation (tree and grass species), about two dozens of various roof shingles, a similar number of sidings and various paving and building materials (bricks of different brands and colors, stained woods, vinyl and several types of painted metals), and car paints. Many of these materials are pointed out in Fig. 4. This image was clustered from mU-matrix visualization in [33], as shown in Fig. 4 and 5.

4.2   Clustering of the Ocean City Multispectral Image

We show clustering with CONNvis on an image of Ocean City, described in Sect. 4.1. Fig. 11 compares an earlier benchmark supervised classification (24 classes, [18]) with an SOM clustering obtained through CONNvis as described in Fig. 10. Fig. 12 shows the extracted 27 clusters on the SOM and an enlarged part of the CONNvis for details of cluster separations including several small clusters. The two thematic maps in Fig. 11 have a strong similarity, which suggests that the clustering found all supervised classes. In addition, it discovered several new ones (a, c, j), which were not known at the time of the supervised classification. From comparison with aerial photographs these appear to be roofs (clusters "a" and "c" in red and blue ovals), and a different surface paving (cluster "j", in white oval) at the end of one road. These and other new units (e, m, subclusters of supervised class M) are distinct enough spectrally to justify separate clusters, as seen in Fig. 13. Another important improvement in the CONNvis clustering is that it assigned labels to many more pixels than the earlier supervised classification, which manifests in more green (vegetation) and turquoise (ocean water) pixels. This is not only because of the discovery of new material units (which are very small) but mostly because the CONNvis view helps quite precise delineation of the cluster boundaries (as shown in Fig. 12).

Comparison with the popular ISODATA clustering is interesting. ISODATA clustering, done in ENVI (ITT Industries, Inc., http://www.ittvis.com/index. asp), was graciously provided by Prof. Bea Csathó of the University of Buffalo [54]. The ISODATA is a refined k-means clustering [59] that has the flexibility to iteratively come up with an optimum number of clusters capped at a user specified maximum. Using the default parameters values in ENVI for a maximum of five iterations, 10 clusters, shown in Fig. 14, left, were produced when up to 10 cluster centers were allowed. The 18 clusters in Fig. 14, right, resulted for a maximum of 20 (and also for 30) clusters. Experiments allowing more iterations and more clusters to be merged produced no visible change. To help visual comparison with the SOM maps, we tried to recolor the randomly assigned ISODATA colors to those in the cluster map in Fig. 11, by assigning to each ISODATA cluster the color of that SOM cluster which is most frequent in the given ISODATA cluster. (This obviously has limits since clusters formed by two different algorithms are not necessarily the same. For the same reason the color wedges and labels of each cluster map are different.) For example, the ISODATA cluster G was assigned the color of SOM cluster G (road, concrete) since the road surfaces dominate that ISODATA cluster. This recoloring immediately shows that in the 10-cluster case (Fig 14, left) this ISODATA cluster also comprises several roof types (SOM clusters B, C, D, E, F, U and V) that are spectrally distinct and resolved in the SOM map. ISODATA formed superclusters of the 27 clusters in the SOM map. Similar supergroups can be seen for the vegetation.

The 18-cluster case (Fig 14, right) is more complicated, but ISODATA still formed recognizable superclusters. The correspondence between water bodies is obvious. SOM clusters G (road) and B now have one-to-one match with the same labels in the ISODATA map, but there is also confusion among clusters. For example, B (orchid color, concrete roof), in the ISODATA map also includes SOM clusters E (the divider paint and a roof type, light blue) and M, a bare lot (yellow, at the top row of houses), in spite that E and M are distinct spectrally.

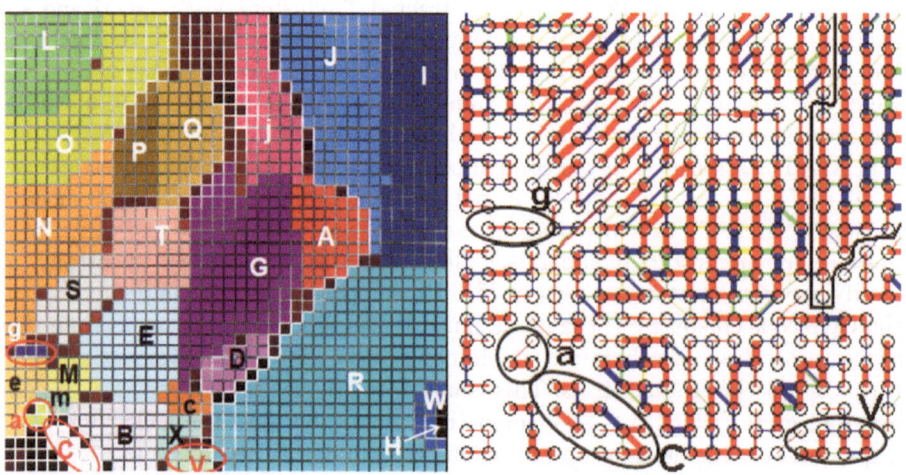

Fig. 12. Cluster extraction for the Ocean City data from CONNvis. **Left:** The extracted clusters in the SOM shown with the same color codes as in Fig. 11, right. **Right:** CONNvis of the bottom left quarter of the SOM to illuminate the representation of cluster boundaries. The small clusters C, g, V and a are clearly separated.

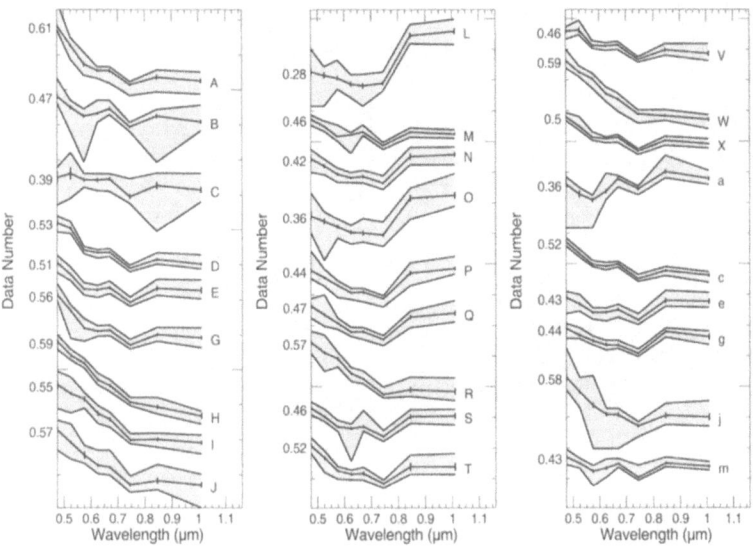

Fig. 13. Spectral statistics of the 27 SOM / CONNvis clusters of the Ocean City image (Fig. 11, right). Mean spectra with standard deviations (vertical bars) and the envelope of each cluster (shaded area) are displayed, vertically offset for viewing convenience. The number at left of each curve indicates the DN value in the first channel. Most of the large clusters (A, B, E, G, I, J, L, N, O, P, Q, R, S, T, j) are tight, suggesting clean delineation of boundaries.

The spectral plots in Fig. 13 and 15 also give partial indication of this confusion, relative to SOM clustering. It is obvious that in both cases the spectral clusters have overlaps (as expected). There is, however, sufficient discriminating information in the non-overlapping bands that the SOM / CONNvis clustering was able to utilize. Most of the 27 CONNvis clusters are reasonably tight (their envelope follows the mean, and standard deviations are small). 6–8 (about one fourth) of the clusters appear to have outliers indicated by loose envelopes but still small standard deviations. In comparison, half of the 18 ISODATA clusters have loose envelopes. More interestingly, most of the large clusters (see listing in the caption of Fig. 13) are tight in the CONNvis plots whereas most of the large clusters (listed in the caption of Fig. 15) are loose in the ISODATA case. This suggests that the boundary delineations by CONNvis, which are in agreement with the benchmark classification, are cleaner. ISODATA forces spherical clusters, whose boundaries may significantly differ from the natural cluster boundaries.

As shown in [53] for a 196-band hyperspectral image of another part of Ocean City containing 30 verified clusters, with increased complexity of the data (when also a larger number of clusters were allowed for ISODATA) ISODATA's confusion of the true clusters greatly increased compared to what we show here for the multispectral case.

Fig. 14. ISODATA clusterings of the Ocean City image. **Left:** 10 clusters resulting from allowing 5 to 10 clusters. Examination of details reveals that the ISODATA clustering represents quite clean cut superclusters of the SOM clusters, as discussed in the text. **Right:** 18 clusters resulting from allowing 10 to 20 clusters. The ISODATA clusters still form supergroups of the SOM clusters, but some confusion also occurs. The spectral plots in Fig. 13 and 15 provide an insight to, and comparison of cluster separations.

4.3   Clusterings of the RIT Synthetic Hyperspectral Image

The 210-band RIT image, described in Sect. 4.1 and in Fig. 4, presents a case of extreme variations among a large number of clusters. This diversity is reflected in Fig. 5 where 39 extracted clusters are overlain on the SOM of this data set, and the spatial distribution of the clusters in the image is also shown. The number of data points in a cluster varies from 1 to nearly 40,000. Many clusters (twenty some different roof types of single houses) have only 200–400 pixels, and several makes of cars (clusters c, f, W, e, noticable mostly in the parking lot in the center of the scene) are represented by less than 10 image pixels. These very small clusters each occupy 1–2 prototypes at the upper and lower left corners of the SOM, along with a few groups of 4–6 prototypes (for example U (lilac), B (orchid), k (medium purple), or E (light blue)), which map larger buildings, very apparent in the scene. The spectra of these cover types exhibit a wide range of similarities. For example, the paving (cluster V, light green), and the grass (K, pure green) are very different, indicated by the strong mU-matrix fences; while a subset of the asphalt shingle roofs have quite subtle yet consistent differences [33]. (As explained in Sect. 3.1 color limitations restricted us to show only about half of the more than 70 clusters. We also used a common color code for several vegetation types, dark blue clusters in the lower right of the SOM, to make color variations more effective for the small clusters which are of greater interest.)

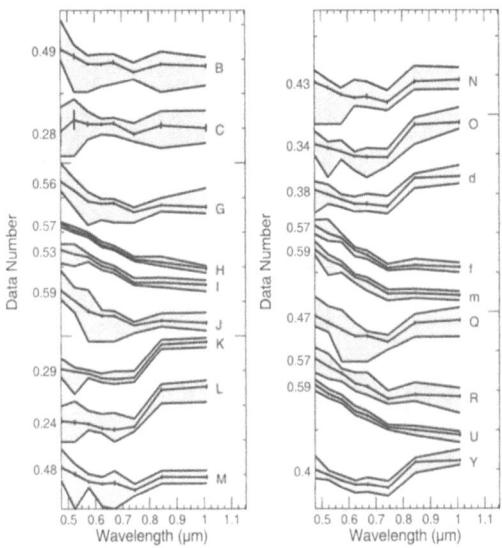

Fig. 15. Spectral statistics of the 18 ISODATA clusters of the Ocean City image (Fig. 14, right). Mean spectra with standard deviations (vertical bars) and the envelope of each cluster (shaded area) are displayed, vertically offset for viewing convenience. The number at left of each curve indicates the DN value in the first channel. Most of the large clusters (B, G, I, J, K, L, M, N, O. Q, R) appear loose.

The details of this clustering from a mU-matrix representation, including descriptions of the surface materials (cover types), spectral characterization showing similarities and differences, matches with the known true classes, and demonstration of discovery of various cars (tiny clusters), are published in [33]. Since we can capture more details with either mU-matrix or CONNvis than with ISODATA (as shown in Sect. 4.2 and in [53]), here we want to examine the relative merits of mU-matrix and CONNvis representations for extraction of clusters from this complicated data set.

We show two clusterings from two SOMs in Fig. 16, which were learned separately but with the same parameters and both to 3M steps. Consequently they are very similar with some minor differences, thus we can make comparative observations between these two clusterings. The top row of Fig. 16 presents the one from mU-matrix visualization in [33], the bottom row shows one that resulted from CONNvis clustering. (Owing to random assignments the same clusters have different colors in the two maps, but the very similar layout helps relate them visually. The cluster labels we use in this section refer to the key in the color wedge in Fig. 5.) The mU-matrix fences are also superimposed. In addition, empty prototypes are shown as medium grey cells (mostly at the boundaries of major clusters). Empty prototypes can be overlain by cluster labels (colors) to produce a homogeneous look of a cluster such as in the case of most of the large clusters here. We removed the color label of cluster V (light green in the top

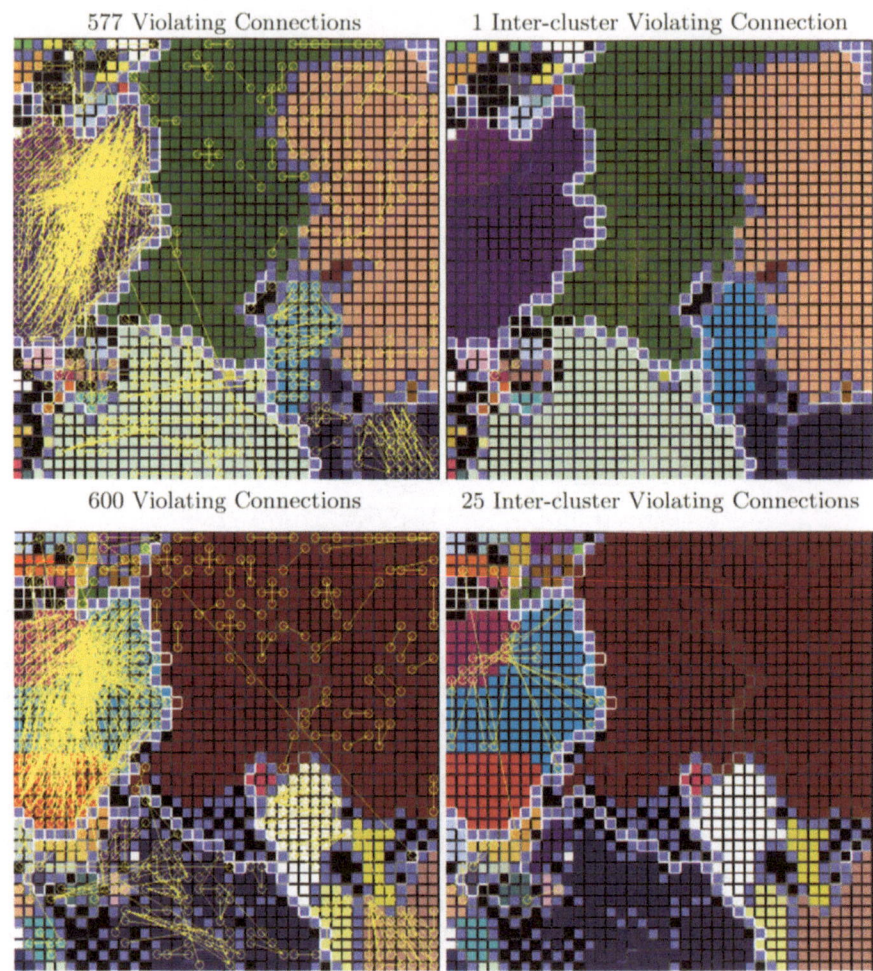

Fig. 16. Comparison of two clusterings of the RIT data. **Top** row: clusters extracted from mU-matrix knowledge, and color coded as in Fig. 5. **Bottom** row: clusters captured from CONNvis. Color coding of clusters is different from the top row because of random label assignments. The mU-matrix fences are superimposed on both. Medium grey cells are empty prototypes, whereas black cells are prototypes left unclustered for reasons of color limitation. **Left**: TopoView visualization (yellow lines) of all violating connections are overlain. **Right**: inter-cluster violating connections are shown.

row) and from the lower part of the large cluster K (pure green) in the CONNvis clustering (bottom row) to show the underlying scattered empty prototypes. We overlaid, on top of the extracted clusters, the TopoView showing all violating connections as yellow lines on the left, and inter-cluster violating connections on the right (i.e., violating connections which have two ends in the same cluster, or either end is an empty or unlabelled prototype were omitted). TopoView confirms that the two SOMs are very similarly organized. One significant detail is that in both maps there are few inter-cluster violating connections going into or out of the small clusters in the upper and lower left corners.

The first impression is that the two clusterings are very similar. The most striking difference involves the large clusters K (pure green, grass) and T (flesh color, grass/dirt mix) at the center and right of the SOM at top, which were extracted as one cluster (largest, rust color area) in the SOM at the bottom. At top, a corridor of empty protoypes is clearly visible between the clusters K and T. In the mU-matrix representation, this corridor is accompanied by a double fence of consistent height, indicating a uniform difference between K and T along this corridor. (The absolute height of this fence is lower than that of the other, more prominent ones in this figure, but it can be seen unmistakably by interactive setting of the visualization range for fence values.) Similar corridor exists under the large rust color cluster in the bottom. However, many other empty prototypes are also present in both SOMs under these clusters, mostly in the checkered layout as seen at the unclustered part of the bottom SOM (or in Fig. 7). In the CONNvis these are all unconnected (no connection goes outside of the manifold, not shown here), therefore the empty cells could outline cluster boundaries the same way as the unconnected prototypes do in Fig. 12 for the Ocean City data. The difference is that in the case of Ocean City a corridor was cut by severing similarly weak connections in a contiguous area of non-empty prototypes, whereas here there are no connections to evaluate, the discontinuities have equal importance and therefore the same corridor does not emerge from the checkered pattern. As a result CONNvis sees these two units as a field of many small clusters. Since CONNvis lacks the distance information from which to notice that some of the discontinuities caused by these small clusters are "more discontinuous" than others, one cannot infer that there are two groups of small clusters in this field between which the (distance based) spectral dissimilarity is much greater than the dissimilarities within each of these two groups. The same scattered empty prototypes may be the cause of some small clusters not being apparent in the CONNvis, for example the tennis court, which is mapped to one prototype in the upper SOM (O, split pea color), between the light green cluster V and cluster K; or the running paths of the baseball field (cluster Y, rust color, wedged into cluster T in the top SOM). Under these circumstances CONNvis may not have the tool to distinguish those single-prototype clusters that are usually obvious from their "fenced off" appearance in a mU-matrix.

It is interesting to note that the relatively large cluster "g" (purple, at center left of the top SOM) is the most disorganized according to TopoView: it has the most violating connections. The reason is that this cluster is extremely noisy. In contrast, the largest clusters seem well organized with only spurious topology violations. However, the boundaries of all clusters including "g" were cleanly extracted, which is indicated by the lack of inter-cluster violations at top right. These connections were not thresholded by strength, the few showing in the bottom right SOM are mostly weak and unimportant.

TopoView reveals another important difference between the two clusterings: the splitting of the large cluster "g" into two (turqoise and red) by CONNvis. These two clusters are much more similar to one another in their spectral statistics (mean, standard deviation, not shown here), i.e., by distance based similarity, than the clusters K and T discussed above. Yet, CONNvis separates them by connectivity measure. In contrast to the clusters K and L, here the SOM has a contiguous field of non-empty prototypes, thus the relative connectivity strengths can be evaluated and cuts made as prescribed for CONNvis cluster extraction (Fig. 10). These two subclusters were not visible from the distance based (mU-matrix) representation, where the entire parent clsuter "g" appears to have a fairly uniform mesh of high fences.

Concluding from these discussions, the mU-matrix can be difficult to interpret where prototype distances may be very similar but relatively large within clusters with large variance. In contrast, CONNvis can be blind to distance based similarities. This susggests that alternating or using these two representations together would further increase the effectiveness of cluster extraction.

5 Conclusions and Outlook

We concentrate on issues related to cluster identification in complicated data structures with SOMs, including the assessment and monitoring of the topology preservation of the mapping. We distinguish a level of order in the SOM that is acceptable for cluster extraction. This can be achieved much earlier in the learning than finely tuned topography matching, but not as early as a sorted state. A sorted state, the mapping of known true classes in the SOM without scrambling does not guarantee successful detection of the same entities, because the prototypes may still not be molded sufficiently for a mU-matrix or other distance based similarity representations to align with the natural cluster boundaries. (Fig. 3, top center, is an example.) Our tool TopoView, presented in Sect. 3.1 can serve for the assessment of topology preservation on this level in relation to mU-matrix knowledge, as well as a verification tool for extracted clusters. The CONNvis SOM visualization, also a recent development, and our long time tool, the modified U-matrix, help achieve very detailed extraction of many relevant clusters, as shown in Sect. 4.2 and 4.3, representing dramatic improvement over some existing popular clustering capabilities such as ISODATA, for highly structured manifolds.

However, we point out that our tools could be further improved by combining the distance based knowledge of the mU-matrix and the topology based knowledge of the CONNvis. A natural extension will be to combine these two into one similarity measure, based on our experiences.

We do not discuss some aspects which could significantly contribute to SOM clustering but have not been much researched. Map magnification is one. This interesting subject is explored in [18] for highly structured data. Methods for verification of clusters (extracted by any algorithm) against the natural partitions in the data, are generally lacking for complicated data. Existing cluster validity indices, which tend to favor particular types of data distributions (such as spherical clusters) fail to give accurate evaluation of the clustering quality for highly structured data. This is discussed in [60] and a new validity index, based on the same connectivity (CONN) matrix as used in CONNvis, is offered.

A valuable aspect of CONNvis SOM clustering is that it seems amenable to automation. Since the binning of connectivity stregths (line widths) in Fig. 9 is generated with thresholds derived automatically from the data characteristics as defined in [34], these thresholds can provide meaningful guidance for finding thinly textured parts of the manifold and cutting connections to achieve cluster separation. For this to work, however, the designation of coarse clusters by the human operator (as in Fig. 10) will need to be replaced by an automated consideration of the relationships between local and global connectedness at each prototype. While this is non-trivial we think it is doable and we are gathering insights from interactive CONNvis clustering for how to best implement this. Successful automation, with the same level of sophistication as shown here for interactive clustering, would significantly contribute to the solution of large problems such as on-board autonomous science, detection of small targets from unmanned vehicles in war zones, or precise mining of large security data bases.

Acknowledgments

We thank Drs. Juha Vesanto and Esa Alhoniemi for sharing their "Clown" data set and the SOM weights from their processing in [52]; Prof. John Kerekes, Rochester Intstitute of Technology, for providing the synthetic hyperspectral image used for this work, and Prof. Maj. Michael Mendenhall of the Air Force Institute of Technology, United States Air Force, for his help with preprocessing the same; Prof. Bea Csathó, University of Buffalo, for the Ocean City data and ground truth, as well as for ISODATA clusterings of the same. This work was partially supported by the Applied Information Systems Research Program (grant NNG05GA94G) of NASA's Science Mission Directorate.

References

1. Lee, J., Verleysen, M.: Nonlinear Dimensionality Reduction. Information Science and Statistics. Springer, New York (2007)
2. Gorban, A., Kégl, B., Wunsch, D., Zinovyev, A. (eds.): Principal Manifolds for data Visualization and Dimension Reduction. Lecture Notes in Computational Science and Engineering. Springer, New York (2008)
3. Cox, T.F., Cox, M.: Multidimensional Scaling. Chapman and Hall/CRC, Boca Raton (2001)
4. Tenenbaum, J.B., de Silva, V., Langford, J.: A global geometric framework for nonlinear dimensionality reduction. Science 290(5500), 2319–2323 (2000)
5. Roweis, S., Soul, L.: Nonlinear dimensionality reduction by locally linear embedding. Science 290(5500), 2323–2326 (2000)
6. Donoho, D.L., Grimes, C.: Hessian eigenmaps: new locally linear embedding techniques for high-dimensional data. Proc. National Academy of Sciences. 100, 5591–5596 (2003)

7. Pless, R.: Using Isomap to explore video sequences. In: Proc. International Conference on Computer Vision, pp. 1433–1440 (2003)
8. Yang, M.: Face Recognition Using Extended Isomap. In: Proc. International Conference on Image Processing ICIP 2002, vol. 2, pp. 117–120 (2002)
9. Polito, M., Perona, P.: Grouping and dimensionality reduction by locally linear embedding. In: Proc. Neural Information Processing Systems, NIPS (2001)
10. Vlachos, M., Domeniconi, C., Gunopulos, D., Kollios, G., Koudas, N.: Non-linear dimensionality reduction techniques for classification and visualization. In: Proceedings of 8th SIGKDD, pp. 645–651 (2002)
11. Zhang, J., Li, S.Z., Wang, J.: Manifold learning and applications in recognition. In: Tan, Y.P., Kim Hui Yap, L.W. (eds.) Intelligent Multimedia Processing with Soft Computing. Springer, Heidelberg (2004)
12. Kohonen, T.: Self-Organizing Maps, 2nd edn. Springer, Heidelberg (1997)
13. Martinetz, T., Berkovich, S., Schulten, K.: Neural Gas network for vector quantization and its application to time-series prediction. IEEE Trans. on Neural Networks 4(4), 558–569 (1993)
14. Cottrell, M., Hammer, B., Hasenfuss, A., Villmann, T.: Batch and median neural gas. Neural Networks 19, 762–771 (2006)
15. Bishop, C.M., Svensen, M., Williams, C.K.I.: GTM: The Generative Topographic Mapping. Neural Computation 10(1), 215–234 (1998)
16. Aupetit, M.: Learning topology with the Generative Gaussian Graph and the EM Algorithm. In: Weiss, Y., Schölkopf, B., Platt, J. (eds.) Advances in Neural Information Processing Systems 18, pp. 83–90. MIT Press, Cambridge (2006)
17. Bauer, H.U., Der, R., Herrmann, M.: Controlling the magnification factor of self–organizing feature maps. Neural Computation 8(4), 757–771 (1996)
18. Merényi, E., Jain, A., Villmann, T.: Explicit magnification control of self-organizing maps for "forbidden" data. IEEE Trans. on Neural Networks 18(3), 786–797 (2007)
19. Villmann, T., Claussen, J.: Magnification control in self-organizing maps and neural gas. Neural Computation 18, 446–469 (2006)
20. Hammer, B., Hasenfuss, A., Villmann, T.: Magnification control for batch neural gas. Neurocomputing 70, 1125–1234 (2007)
21. DeSieno, D.: Adding a conscience to competitive learning. In: Proc. IEEE Int'l Conference on Neural Networks (ICNN), New York, July 1988, vol. I, pp. I–117–124 (1988)
22. Cottrell, M., Fort, J., Pages, G.: Theoretical aspects of the SOM algorithm. Neurocomputing 21, 119–138 (1998)
23. Ritter, H., Schulten, K.: On the stationary state of Kohonen's self-organizing sensory mapping. Biol. Cyb. 54, 99–106 (1986)
24. Erwin, E., Obermayer, K., Schulten, K.: Self-organizing maps: ordering, convergence properties and energy functions. Biol. Cyb. 67, 47–55 (1992)
25. Hammer, B., Villmann, T.: Mathematical aspects of neural networks. In: Proc. Of European Symposium on Artificial Neural Networks (ESANN 2003), Brussels, Belgium. D facto publications (2003)
26. Martinetz, T., Schulten, K.: Topology representing networks. Neural Networks 7(3), 507–522 (1994)
27. Villmann, T., Der, R., Herrmann, M., Martinetz, T.: Topology Preservation in Self–Organizing Feature Maps: Exact Definition and Measurement. IEEE Transactions on Neural Networks 8(2), 256–266 (1997)
28. Bauer, H.U., Pawelzik, K.: Quantifying the neighborhood preservation of Self-Organizing Feature Maps. IEEE Trans. on Neural Networks 3, 570–579 (1992)

29. Kiviluoto, K.: Topology preservation in self-organizing maps. In: Proceedings IEEE International Conference on Neural Networks, Bruges, June 3–6, 1996, pp. 294–299 (1996)

30. Zhang, L., Merényi, E.: Weighted Differential Topographic Function: A Refinement of the Topographic Function. In: Proc. 14th European Symposium on Artificial Neural Networks (ESANN 2006), Brussels, Belgium, pp. 13–18. D facto publications (2006)

31. Csathó, B., Krabill, W., Lucas, J., Schenk, T.: A multisensor data set of an urban and coastal scene. In: Int'l Archives of Photogrammetry and Remote Sensing, vol. 32, pp. 26–31 (1998)

32. Bodt, E., Verleysen, M.C.: Statistical tools to assess the reliability of self-organizing maps. Neural Networks 15, 967–978 (2002)

33. Merényi, E., Tasdemir, K., Farrand, W.: Intelligent information extraction to aid science decision making in autonomous space exploration. In: Fink, W. (ed.) Proceedings of DSS 2008 SPIE Defense and Security Symposium, Space Exploration Technologies, Orlando, FL, Mach 17–18, 2008, vol. 6960, pp. 17–18. SPIE (2008) 69600M Invited

34. Tasdemir, K., Merényi, E.: Exploiting data topology in visualization and clustering of Self-Organizing Maps. IEEE Trans. on Neural Networks (2008) (in press)

35. Ultsch, A.: Self-organizing neural networks for visualization and classification. In: Opitz, O., Lausen, B. (eds.) Information and Classification — Concepts, Methods and Applications, pp. 307–313. Springer, Berlin (1993)

36. Kraaijveld, M., Mao, J., Jain, A.: A nonlinear projection method based on Kohonen's topology preserving maps. IEEE Trans. on Neural Networks 6(3), 548–559 (1995)

37. Merkl, D., Rauber, A.: Alternative ways for cluster visualization in Self-Organizing Maps. In: Proc. 1st Workshop on Self-Organizing Maps (WSOM 1997), Espoo, Finland, June 4-6, 1997, pp. 106–111 (1997)

38. Ultsch, A.: Maps for the visualization of high-dimensional data spaces. In: Proc. 4th Workshop on Self-Organizing Maps (WSOM 2003), Paris, France, vol. 3, pp. 225–230 (2003)

39. Cottrell, M., de Bodt, E.: A Kohonen map representation to avoid misleading interpretations. In: Proc. 4th European Symposium on Artificial Neural Networks (ESANN 1996), pp. 103–110. D-Facto, Bruges (1996)

40. Himberg, J.: A SOM based cluster visualization and its application for false colouring. In: Proc. IEEE-INNS-ENNS International Joint Conf. on Neural Networks, Como, Italy, vol. 3, pp. 587–592 (2000)

41. Kaski, S., Venna, J., Kohonen, T.: Coloring that reveals cluster structures in multivariate data. Australian Journal of Intelligent Information Processing Systems 6, 82–88 (2000)

42. Villmann, T., Merényi, E.: Extensions and modifications of the Kohonen-SOM and applications in remote sensing image analysis. In: Seiffert, U., Jain, L.C. (eds.) Self-Organizing Maps: Recent Advances and Applications, pp. 121–145. Springer, Heidelberg (2001)

43. Vesanto, J.: SOM-Based Data Visualization Methods. Intelligent Data Analysis 3(2), 111–126 (1999)

44. Kaski, S., Kohonen, T., Venna, J.: Tips for SOM Processing and Colourcoding of Maps. In: Deboeck, G., Kohonen, T. (eds.) Visual Explorations in Finance Using Self-Organizing Maps, London (1998)

45. Pölzlbauer, G., Rauber, A., Dittenbach, M.: Advanced visualization techniques for self-organizing maps with graph-based methods. In: Jun, W., Xiaofeng, L., Zhang, Y. (eds.) Proc. Second Intl. Symp. on Neural Networks (ISSN 2005), Chongqing, China, pp. 75–80. Springer, Heidelberg (2005)
46. Aupetit, M., Catz, T.: High-dimensional labeled data analysis with topology representing graphs. Neurocomputing 63, 139–169 (2005)
47. Aupetit, M.: Visualizing the trustworthiness of a projection. In: Proc. 14th European Symposium on Artificial Neural Networks, ESANN 2006, Bruges, Belgium, April 26-28, 2006, pp. 271–276 (2006)
48. Howell, E.S., Merényi, E., Lebofsky, L.A.: Classification of asteroid spectra using a neural network. Jour. Geophys. Res. 99(E5), 10, 847–10, 865 (1994)
49. Merényi, E., Howell, E.S., et al.: Prediction of water in asteroids from spectral data shortward of 3 microns. ICARUS 129, 421–439 (1997)
50. Tasdemir, K., Merényi, E.: Considering topology in the clustering of self-organizing maps. In: Proc. 5th Workshop On Self-Organizing Maps (WSOM 2005), Paris, France, September 5–8, 2005, pp. 439–446 (2005)
51. Tasdemir, K., Merényi, E.: Data topology visualization for the Self-Organizing Map. In: Proc. 14th European Symposium on Artificial Neural Networks (ESANN 2006), Brussels, Belgium, April 26–28, 2006, pp. 125–130. D facto publications (2006)
52. Vesanto, J., Alhoniemi, E.: Clustering of the self-organizing map. IEEE Transactions on Neural Networks 11(3), 586–600 (2000)
53. Merényi, E., Csató, B., Taşdemir, K.: Knowledge discovery in urban environments from fused multi-dimensional imagery. In: Gamba, P., Crawford, M. (eds.) Proc. IEEE GRSS/ISPRS Joint Workshop on Remote Sensing and Data Fusion over Urban Areas (URBAN 2007), Paris, France, IEEE Catalog number 07EX1577, April 11-13, 2007, pp. 1–13 (2007)
54. Csathó, B., Schenk, T., Lee, D.C., Filin, S.: Inclusion of multispectral data into object recognition. Int'l Archives of Photogrammetry and Remote Sensing 32, 53–61 (1999)
55. Schott, J., Brown, S., Raqueño, R., Gross, H., Robinson, G.: An advanced synthetic image generation model and its application to multi/hyperspectral algorithm development. Canadian Journal of Remote Sensing 25(2) (June 1999)
56. Ientilucci, E., Brown, S.: Advances in wide-area hyperspectral image simulation. In: Proceedings of SPIE, May 5–8, 2003, vol. 5075, pp. 110–121 (2003)
57. Green, R.O.: Summaries of the 6th Annual JPL Airborne Geoscience Workshop, 1. In: AVIRIS Workshop, Pasadena, CA, March 4–6 (1996)
58. Green, R.O., Boardman, J.: Exploration of the relationship between information content and signal-to-noise ratio and spatial resolution. In: Proc. 9th AVIRIS Earth Science and Applications Workshop, Pasadena, CA, February 23–25 (2000)
59. Tou, J., Gonzalez, R.C.: Pattern Recognition Principles. Addison-Wesley Publishing Company, Reading (1974)
60. Tasdemir, K., Merényi, E.: A new cluster validity index for prototype based clustering algorithms based on inter- and intra-cluster density. In: Proc. Int'l Joint Conf. on Neural Networks (IJCNN 2007), Orlando, FL, August 12–17, 2007, pp. 2205–2211 (2007)

Estimation of Boar Sperm Status Using Intracellular Density Distribution in Grey Level Images

Lidia Sánchez[1] and Nicolai Petkov[2]

[1] Department of Mechanical, Computing and Aerospace Engineerings
University of León,
Campus de Vegazana s/n, 24071 León, Spain
[2] Institute of Mathematics and Computing Science
University of Groningen,
P.O. Box 407, 9700 AK Groningen, The Netherlands
lidia.sanchez@unileon.es

Abstract. In this work we review three methods proposed to estimate the fraction of alive sperm cells in boar semen samples. Images of semen samples are acquired, preprocessed and segmented in order to obtain images of single sperm heads. A model of intracellular density distribution characteristic of alive cells is computed by averaging a set of images of cells assumed to be alive by veterinarian experts. We quantify the deviation of the distribution of a cell from this model and use it for classification deploying three different approaches. One is based on a decision criterion used for single cell classification and gives misclassification error of 20.40%. The other two methods are focused on estimating the fraction of alive sperm in a sample, instead of single cell classification. One of them applies the least squares method, achieving an absolute error below 25% for 89% of the considered sample images. The other uses an iterative procedure to find an optimal decision criterion that equalizes the number of misclassifications of alive and dead cells. It provides an estimation of the fraction of alive cells that is within 8% of its actual value for 95% of the samples.

1 Introduction

The quality of boar semen samples used for artificial insemination is analyzed in order to guarantee success in the fertilization process. For this purpose, artificial insemination centers consider a set of characteristics like sperm concentration, motility or morphology in order to assess the percentage of spermatozoa that are potentially fertile [5,15]. Several computer systems have been developed to carry out this analysis automatically [21,14,3]. These systems do not analyze the acrosome status which is another feature that determines the fertilizing capacity of a sperm cell [13]. The acrosome is a cap-like structure in the anterior part of a sperm cell that contains enzymes which are important for reaching and fertilizing an oocyte. If a sperm cell contains a large fraction of cells that have lost their

M. Biehl et al.: (Eds.): Similarity-Based Clustering, LNAI 5400, pp. 169–184, 2009.
© Springer-Verlag Berlin Heidelberg 2009

acrosome, that sample is considered to have low fertilization potential. More information about the role of the acrosome in the fertilization process can be found in [10]. It is also important to estimate the life status of the spermatozoa since a dead cell will never be able to fertilize an oocyte.

There are many approaches to evaluate semen sample images. Some apply Fourier descriptors [7] or wavelets [22] to extract information about the cell morphology and shape. Others study the motility of the spermatozoa, calculating their main direction and velocity of motion [5,9]. Although most works find sub-populations of spermatozoa with a certain motility or shape there is no correlation with the estimation of the fertilization potential made by veterinarians using traditional techniques. For example, in [20] the authors find different sub-populations in a sample that just contains spermatozoa considered as normal by veterinarians.

In previous works, we have proposed several methods to determine the life status of a sperm cell, classifying it automatically as alive or dead according to its intracellular grey level distribution in microscope images [18,17,2,19]. Automatic evaluation of acrosome integrity has been also carried out using techniques based on the gradient of the image [12] or texture descriptors [8,16]. Such estimations are made nowadays by a veterinary expert inspecting visually 100 stained sperm heads by means of an epi-fluorescence microscope. That involves costs due to the deployed materials, the need of qualified personnel and the subjectivity of the procedure dependent on the technician proficiency and the environmental conditions.

In this work, we review the proposed methods to assess the life status of boar spermatozoa using images with no stains. This approach implies a reduction of the costs needed to estimate the sperm quality as compared to traditional techniques based on stains and fluorescence. We classify automatically sperm cells as dead or alive (life status estimation). Our methods are based on a model of the intracellular density characteristic of alive cells. The paper is organized as follows: Section 2 presents procedures deployed to acquire, segment and normalize sperm head images. In Section 3, we explain the model definition. The considered classification methods and their performance are described in Section 4. Finally, conclusions are presented in Section 5.

2   Obtaining Sperm Head Images

2.1   Image Acquisition

Spermatozoa sample images are captured using a digital camera mounted on a phase-contrast microscope (Fig. 1). The used microscope magnification was $40\times$ and the image resolution 1600×1200 pixels. Examples are shown in Fig. 2.

2.2   Preprocessing and Segmentation

In an image of a boar semen sample we can see a number of sperm cells with diverse orientations and tilts (Fig. 3a). It is common to observe overlapping cells

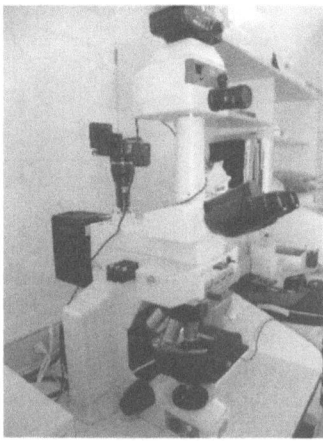

Fig. 1. Microscope with a mounted digital camera used to acquire sperm sample images

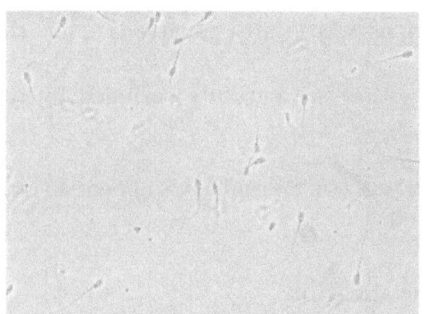 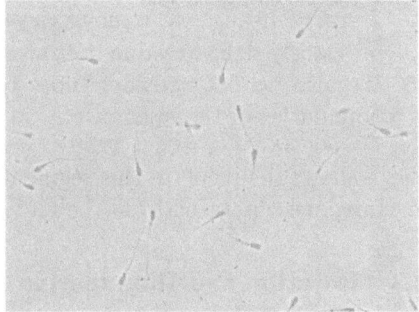

Fig. 2. Examples of boar semen sample images acquired with a digital camera and a phase-contrast microscope

(Fig. 3b), agglutinations of sperm heads (Fig. 3c) and debris (Fig. 3d) due to the manipulation process. The segmentation procedure has to deal with various conditions in a semen sample [4]. In previous works, we have applied several classical border detection algorithms as Sobel, Prewitt, Roberts, Canny, and Isodata. However, the results are not satisfactory [1].

To segment automatically the sperm heads from a boar semen sample image using $40\times$ magnification, we apply the following procedure:

1. We first convert the color images that are captured in positive phase-contrast (see Fig. 2) to grey level images.
2. We apply morphological closing with a disk shaped structuring element to smooth the contours of cells and remove tails.
3. We employ Otsu's method to threshold the image [11] yielding an image with a set of isolated regions that potentially can be sperm heads.

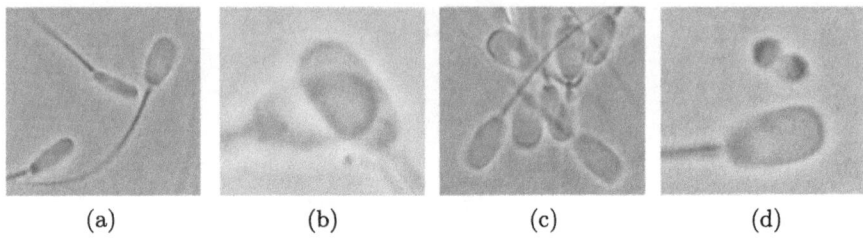

(a)　　　　　　(b)　　　　　　(c)　　　　　　(d)

Fig. 3. Examples of difficulties encountered in segmentation. Boar sperm samples present spermatozoa with (a) different tilts and (b) superimposed with debris or other spermatozoa. For these cases, the intracellular grey level distribution cannot be analyzed. It is also difficult to determine the sperm head borders when (c) there is agglutination of spermatozoa or (d) debris which can be confused with a sperm head.

4. We determine a value *min_area* defined as 45% of the average area of these regions.
5. To avoid analyzing regions that are debris (they usually are smaller than a sperm head), we remove those regions whose area is smaller than the previously defined value *min_area*.
6. We also do not consider those regions that are partially occluded by the boundaries of the image.

Fig. 4 shows the result of this segmentation process. Sperm heads appear as grey level regions separated from the background.

2.3 Rotation and Illumination Normalization

After segmentation, the obtained sperm head images present different orientations and brightness as they are acquired under different illumination conditions.

Sperm heads present oval shapes that can be approximated by ellipses. We compute the major and minor axis of each such ellipse applying principal component analysis [6] (Fig. 5a).

We extract a set of cell boundary points from a segmented sperm head image and apply principal component analysis to this set. The eigenvectors determined in this way serve as major and minor axes of an ellipse that optimally approximates the cell head. Next, we rotate the sperm head image using nearest-neighbor interpolation in order to align the major axis with the horizontal (x) axis (Fig. 5a-b).

Next we consider the horizontal profile of the rotated sperm head image which presents a darker area on the tail side. If necessary, we rotate the image by 180 deg so that this darker side is on the left and the acrosome on the right (Fig. 5b).

A boar sperm head is usually 4 to 5 μm wide and 7 to 10 μm long. In correspondence with these proportions, we re-scale all segmented sperm head

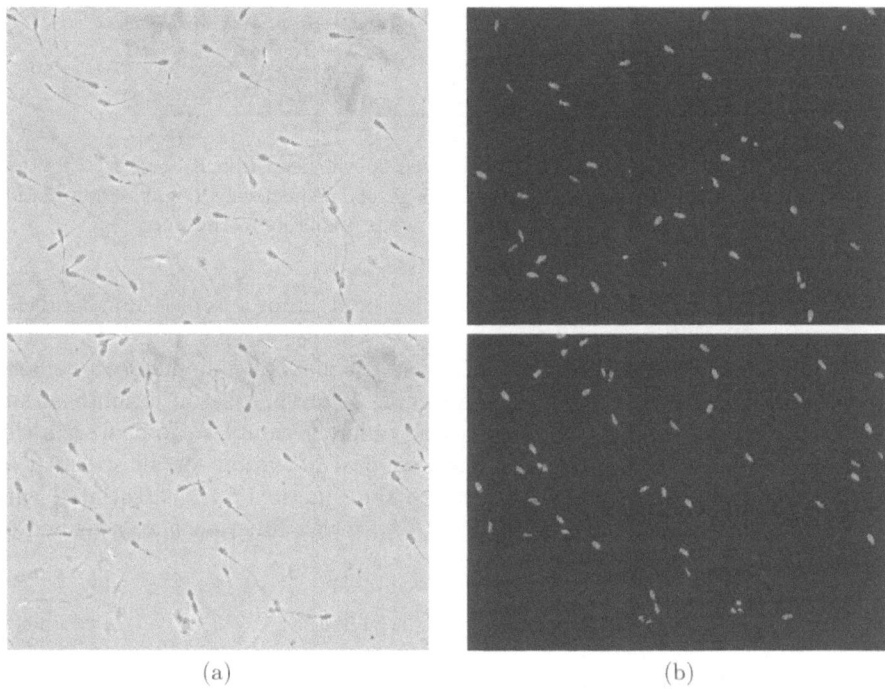

Fig. 4. From the boar semen sample images (left), sperm head images are isolated. The images obtained after the segmentation process (right) present sperm heads as grey level distributions on a black background.

Fig. 5. (a) We find the principal axis of the ellipse that approximates the sperm head. Subsequently, we (b) align the sperm head and (c) re-scale the sperm head image to 19x35 pixels.

images to the same uniform size of 19×35 pixels by applying nearest-neighbour interpolation (Fig. 5c).

Once a sperm head image is aligned and re-sized, we consider a two dimensional function $f(x,y)$ defined by the grey level values of those pixels which belong

Fig. 6. (left) Horizontally aligned and re-sized sperm head image. (center) Elliptical mask with minor and major axis of 19 and 35 pixels, respectively. (right) The function $f(x,y)$ is defined by the grey levels of those pixels which lie in the mask.

to a region S enclosed by an ellipse with minor and major axis of 19 and 35 pixels, respectively (Fig. 6).

To capture the images, the technicians adjust manually the microscope lamp in order to illuminate the boar semen samples. For this reason, brightness and contrast of sperm head images vary across different samples. To deal with this problem, we keep the same mean and standard deviation for all sperm head images by applying a linear transform to the function $f(x,y)$ associated with each image. We transform the function $f(x,y)$ into a function $g(x,y)$ defined on the same support S as follows:

$$g(x,y) = af(x,y) + b \tag{1}$$

where the coefficients a and b are defined as follows:

$$a = \frac{\sigma_g}{\sigma_f}, \ b = \mu_g - a\mu_f \tag{2}$$

The values of the mean μ_f and the standard deviation σ_f of $f(x,y)$ are computed from the sperm head image function f. The values of μ_g and σ_g can be chosen. Empirically, we observe that sperm head images considered in this work present a mean and a standard deviation around 100 and 8, respectively. For this reason, we fix $\mu_g = 100$ and $\sigma_g = 8$.

3   Intracellular Distribution Model

3.1   Feature Extraction

We use a set M of $n = 34$ normalized sperm head images that have been labelled as potentially alive by veterinary experts according to their intracellular distributions $g_i(x,y), i = 1, 2, ...n$ (Fig. 7). Next, we compute a model of the two dimensional intensity distribution $m(x,y)$ of alive cells as a pixel-wise average of these functions (Fig. 8):

$$m(x,y) = \frac{1}{n}\sum_{i=1}^{n} g_i(x,y) \tag{3}$$

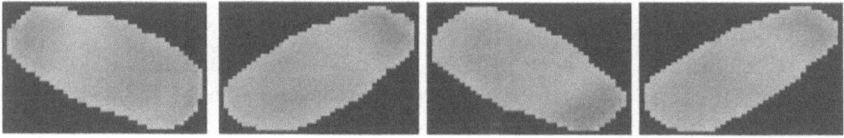

Fig. 7. Examples of sperm head images of the model training set M. Their intracellular distributions are assumed as characteristic of alive spermatozoa by veterinary experts.

Fig. 8. Obtained model of the intracellular distribution for alive sperm head images

For each pixel of the obtained model distribution, we also compute the standard deviation $\sigma(x, y)$ across the images of the model set M:

$$\sigma(x, y) = \sqrt{\sum_{i=1}^{n} \frac{(g_i(x, y) - m(x, y))^2}{n}} \qquad (4)$$

3.2 Estimation of the Deviation from the Model Distribution

We define the dissimilarity d of the distribution function $g(x,y)$ of a given sperm head image to the model distribution function $m(x,y)$, as follows:

$$d = \max \left(\frac{|g(x, y) - m(x, y)|}{\sigma(x, y)} \right) \qquad (5)$$

We consider two training sets A of 44 images and D of 82 images formed by intracellular distributions of sperm head images visually classified by veterinary experts as alive (Fig. 7) or dead (Fig. 9), respectively. We compute the deviations

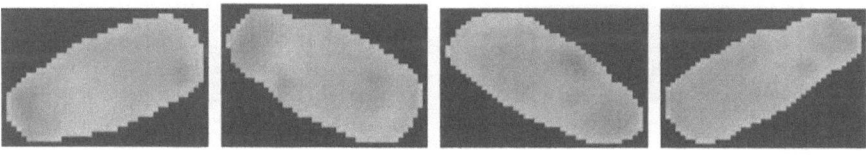

Fig. 9. Examples of sperm heads that were classified by an expert as dead

of these intensity functions from the model distribution using the dissimilarity measure defined above.

Fig. 10 shows the difference between the model distribution function $m(x,y)$ and a few functions $g(x,y)$ that correspond to cells believed to be alive. The difference at a given pixel has been normalized dividing it with the standard deviation at that pixel. Fig. 11 illustrates the same information for cells believed to be dead.

The obtained deviations from the model for the distribution functions of both training sets produce two histograms (Fig. 12).

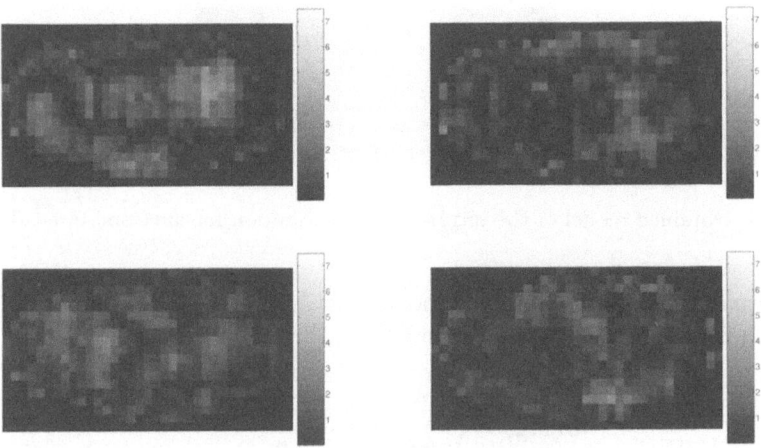

Fig. 10. Examples of the difference between distribution functions of alive cells and the model function

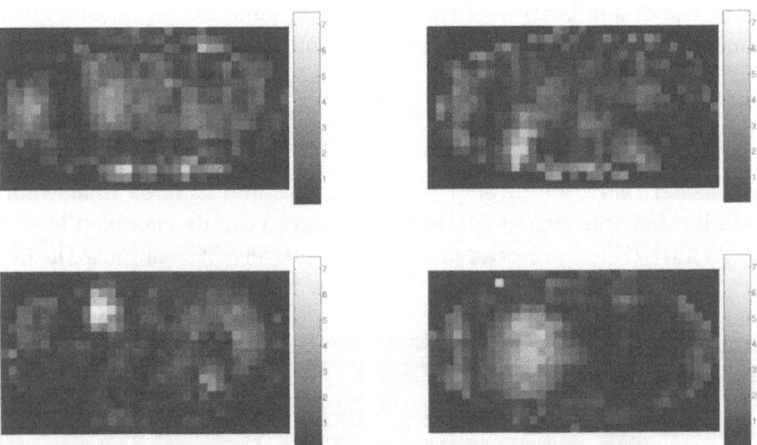

Fig. 11. Examples of the difference between distribution functions of dead cells and the model function

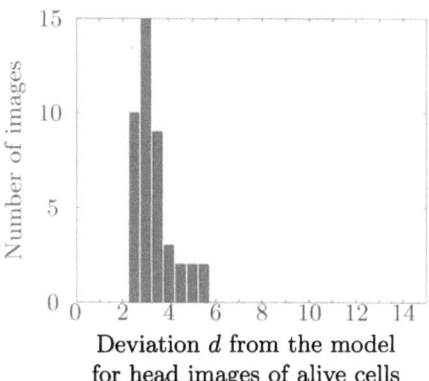
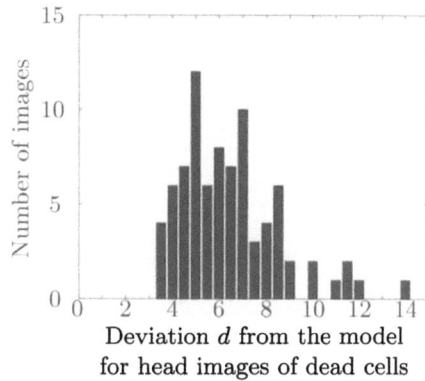

Fig. 12. Histograms defined by the sperm head images of the training sets of alive (left) and dead (right) spermatozoa

4 Classification and Results

Next we propose three classification methods to determine if a sperm cell is alive or dead.

4.1 Minimization of False Acceptance and False Rejection Errors

We choose a classification criterion d_c and classify sperm head images which produce values d of deviation from the model distribution lower than the criterion $(d \leq d_c)$ as alive, otherwise $(d > d_c)$ as dead [17].

While the values of the deviation from the model are below 5.2 for alive cells, the corresponding values for dead cells fall in the range [3, 15]). Since the two histograms overlap in the range (3.5,5.5), it is not possible to select a value of the decision criterion that would lead to error-free classification.

Fig. 13 shows the false acceptance and false rejection errors for the different values of the decision criterion as well as the sum of both errors. The sum of the errors presents a minimum for the value of the decision criterion 4.25.

For validation we consider a test set of 1400 sperm head images labelled as alive or dead. There are 775 images that belong to the alive class and 625 to the dead class. For each image we compute the deviation d from the defined model distribution. A cell is classified as alive if the obtained deviation is smaller than 4.25. Otherwise, it is classified as dead. The achieved false rejection error defined as the percentage of alive sperm head images misclassified as dead is 13.68%. The false acceptance rate defined as the fraction of dead spermatozoa misclassified as alive is 6.72%. So, the overall classification error of this approach is 20.40%. Note that this estimation concerns single cell classification. However, if we consider a boar semen sample that contains alive and dead spermatozoa, and we want to estimate the fraction of alive spermatozoa, false rejection and acceptance errors compensate each other, and the estimation error is equal to the difference of false rejection and false acceptance error, or 6.96%.

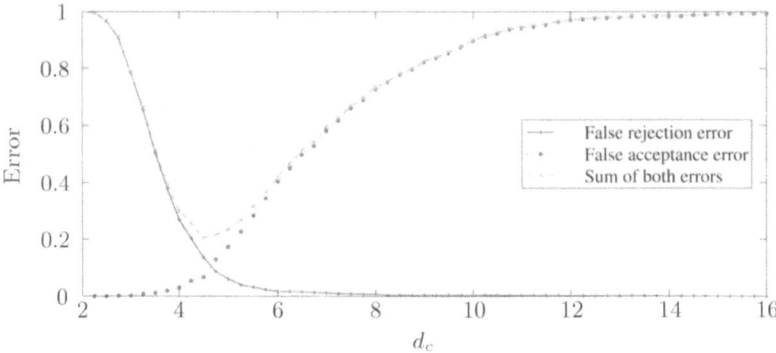

Fig. 13. Error rates in sperm head classification as functions of the decision criterion. The continuous line with the + symbol shows the percentage of alive cells that are misclassified as dead (false rejection error); the dashed line with the ○ symbol represents the fraction of dead cells misclassified as alive (false acceptance error). The dotted line with the × symbol is the sum of the two partial errors and it has a minimum for a decision criterion value of 4.25.

4.2   Equalization of False Rejection and False Acceptance Errors

Next we extend the above method to determine the value of the decision criterion for single cell classification in such a way that the error of the estimated percentage of alive spermatozoa in a boar sample is minimal [19].

We use two sets of sperm head images normalized as explained in Section 2.3. Set A is formed by 718 sperm head images of alive spermatozoa and set D is formed by 650 images of dead sperm cells according to the visual classification made by veterinary experts (Figs. 7 and 9, respectively). The deviations of the normalized distribution functions $g(x,y)$ of these sperm head images from the model intracellular distribution $m(x,y)$ produce two histograms that we consider as approximations of the corresponding class conditional probabilities. Fig. 14 shows the distributions of the occurrence frequencies of the deviation values obtained for the sets A and D.

Since the two distributions overlap, it is not possible to carry out single cell classification without error. Fig. 15 shows the false acceptance $e_a(d_c)$ and false rejection $e_r(d_c)$ errors as functions of the decision criterion d_c.

Our goal is to determine the fraction p of alive spermatozoa in a boar semen sample instead of assessing the life status of each individual spermatozoon. If we deploy single sperm head image classification to estimate the percentage of alive sperm cells in a sample, the fractions of misclassified alive and dead spermatozoa are $pe_r(d_c)$ and $(1-p)e_a(d_c)$, respectively. These errors partially compensate each other because, although some alive spermatozoa are misclassified as dead, there are also dead spermatozoa misclassified as alive. For this reason, the estimation error $e(d_c,p)$ of p is the following function of p and the decision criterion d_c:

$$e(d_c,p) = |pe_r(d_c) - (1-p)e_a(d_c)| \tag{6}$$

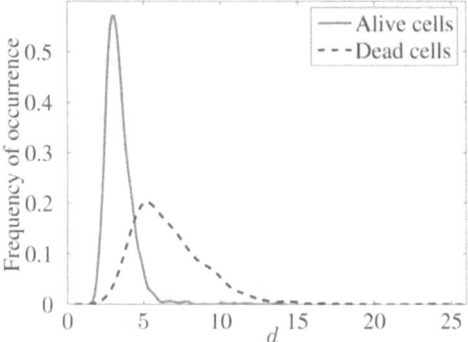

Fig. 14. Probability densities of alive and dead sperm head images as functions of the deviation d from the model distribution of an alive spermatozoon

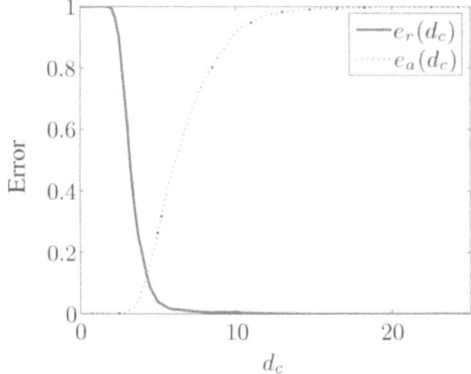

Fig. 15. Misclassification errors $e_a(d_c)$ and $e_r(d_c)$ as functions of the decision criterion d_c which is used to classify sperm head images as alive if $d \leq d_c$ or dead if $d > d_c$

Fig. 16 shows the error $e(d_c, p)$ in the estimation of the fraction of alive spermatozoa in a semen sample as a function of the decision criterion d_c for three different values of p (semen samples with 25%, 50% and 75% of alive spermatozoa).

For each value of p, the error $e(d_c, p)$ in the estimation presents a minimum for a certain value of the decision criterion $d_c(p)$. For that value, the number of misclassified alive spermatozoa is the same as the number of misclassified dead spermatozoa, reducing to zero the error in the fraction estimation $e(d_c, p)$. The optimal value of the decision criterion $d_c(p)$ as a function of p is shown in Fig. 17.

Our goal is to determine the unknown fraction p of alive spermatozoa in a sample. To achieve this with minimal error, we need to know the optimal value of the decision criterion $d_c(p)$ that depends on the unknown p. We propose the following iterative procedure: First, we assume $p_0 = 0.5$ and carry out single

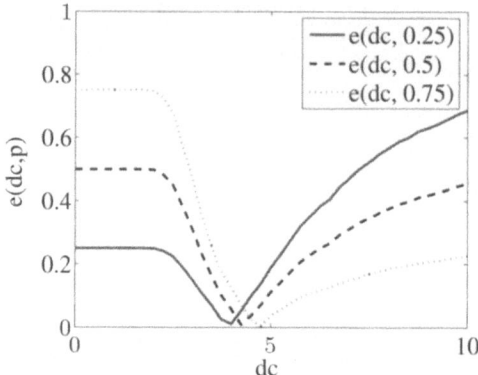

Fig. 16. Error $e(d_c, p)$ in the estimation of the fraction p of alive spermatozoa in a very large sample as a function of the decision criterion d_c for three different values of p

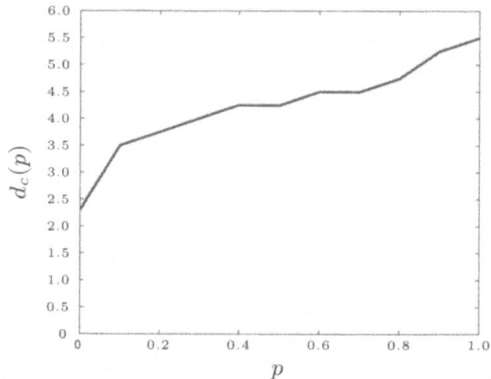

Fig. 17. Value $d_c(p)$ of the decision criterion for which the error in estimating the fraction p of alive spermatozoa in a very large sample is minimal (0)

sperm head image classification. As a result, a fraction p_1 of alive spermatozoa in the sample is estimated. Next, each single sperm head image is classified again but now the decision criterion used is $d_c(p_1)$. This classification gives a new estimation p_2 of the fraction p of alive spermatozoa. The procedure is repeated as often as necessary, whereby the fraction p_i determined in a given step is used to compute the value of the decision criterion $d_c(p_i)$ needed for the next step. In practice, we found that the consecutive estimations of p converge to a stable value after only a few iterations, typically less than 5.

To analyze the error generated with this procedure in the estimation of the fraction of alive spermatozoa, we take a sample of 100 sperm head images. This set comprises $p100$ sperm head images of the set A of alive spermatozoa and $(1-p)100$ images of the set D of dead spermatozoa. We estimate the fraction of alive spermatozoa for that sample following the proposed iterative procedure and

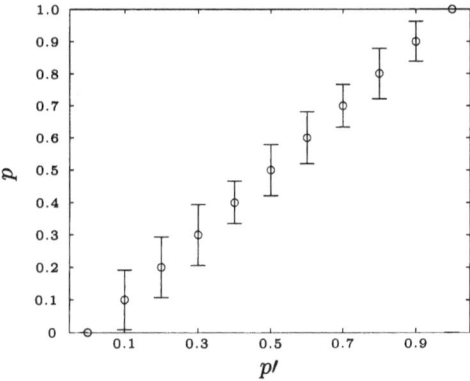

Fig. 18. 95% confidence interval of the fraction of alive spermatozoa p as a function of the estimation $p\prime$

we obtain an estimated fraction $p\prime$. The estimation error of the fraction is $p - p\prime$ which is different for each sample. Therefore, we quantify the fraction estimation error for finite samples of 100 sperm head images by the standard deviation of $p - p\prime$ for 100 samples. In Fig. 18, we represent the 95% confidence interval of the actual fraction of alive spermatozoa as a function of the estimated value. The standard deviation of $p - p\prime$ is smaller than 0.047, being between 0.031 and 0.040 for values $p \geq 0.4$ (a sample used for artificial insemination rarely contains less than 40% of alive spermatozoa). That means that in 95% of the cases the real value of the fraction p of alive spermatozoa in a sample will be within 8% of the estimation $p\prime$ made according to the proposed method.

4.3   Least Squares Estimation of the Fraction of Alive Cells

The histograms shown in Fig. 12 can be considered as class conditional probability distributions $p(d|a)$ and $p(d|de)$ for observing a value d given alive or dead spermatozoon [18], respectively (Fig. 19).

As the goal is to estimate the fraction of alive spermatozoa in a boar semen sample, we can estimate this fraction without carrying out a single classification of each sperm head image. The probability $P(d)$ of observing a value d is given by:

$$P(d) = p(d|a)p + p(d|de)(1 - p) \tag{7}$$

where p and $1 - p$ are the prior probabilities that a given sperm head is alive or dead respectively. We consider $p(d|a)$, $p(d|de)$ and $P(d)$ as known and p as unknown. For different values of d, this equation defines an overdetermined system of linear equations for p that has the following approximate solution given by the least squares method:

$$p = \frac{\sum_d (P(d) - p(d|de))(p(d|a) - p(d|de))}{\sum_d (p(d|a) - p(d|de))^2} \tag{8}$$

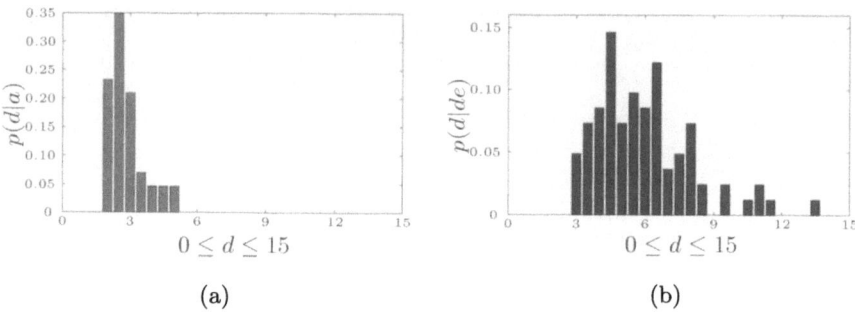

Fig. 19. Class conditional probability distributions $p(d|a)$ and $p(d|de)$ defined by the histograms of the obtained deviations from the model distribution for the training set of alive (a) and dead (b) sperm head images

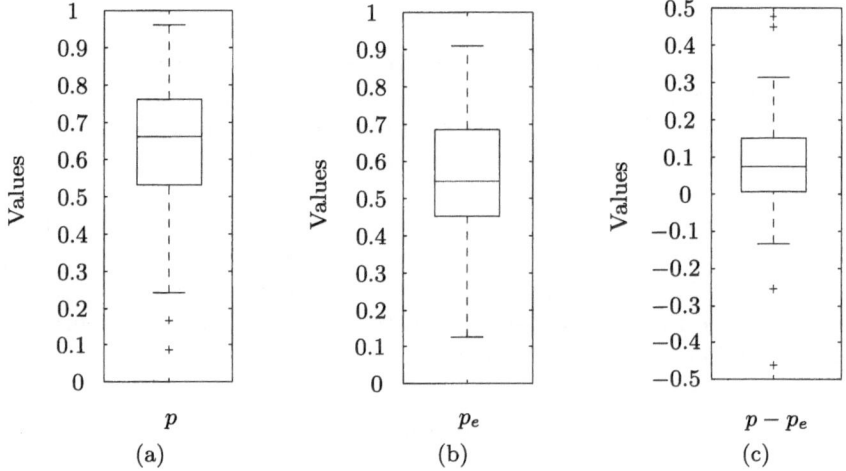

Fig. 20. Box-and whisker diagrams of the values of the fraction of alive sperm heads determined (a) by means of the proposed method, (b) by veterinary experts and (c) of the absolute error of the method in comparison with the experts

Using this formula we can estimate the fraction of alive spermatozoa in a boar semen sample p without classifying each single sperm head image as alive or dead.

We consider a test set of 100 images that contain a variable number of alive and dead spermatozoa (Fig. 2) and obtain a histogram $P(d)$ and a value p for each image. The 100 values of p obtained in this way are illustrated by the box-and-whisker diagram of the Fig. 20a.

The fraction of alive cells estimated by veterinary experts p_e is shown in Fig. 20b. Fig. 20c shows a box-and-whisker diagram of the values of the error $p - p_e$. This error is below 0.25 in 89 of the 100 test images. For 8 further test images, the error is between 0.25 and 0.32, and it is smaller than 0.47 for the 3 remaining images.

5   Summary

We presented an overview of three methods for estimation of the fraction of alive spermatozoa in a boar semen sample. These methods are based on the intracellular distribution of sperm cells in grey level images. The methods are based on considering an intracellular density distribution characteristic of alive spermatozoa and the deviation of the distribution of a given sperm cell from such a model distribution. We use single cell classification or estimation of the fraction of alive sperm cells. In single cell classification we achieve an overall misclassification error of 20.40% of which 6.72% are due to dead and 13.68% to alive cells.

For estimating the fraction of alive spermatozoa, we propose two methods. One of them is based on least square estimation and achieves an absolute error below 25% for 89% of the considered sample images. The other method finds iteratively the fraction of alive spermatozoa and the corresponding optimal decision criterion. It provides an estimation of the fraction of alive cells which is within 8% the real fraction value for 95% of the samples.

References

1. Alegre, E., Fernández, R.A., Sánchez, L., Rodríguez, V., González, R., Tejerina, F., Domínguez, J.C.: Digital image segmentation methods for automatic quality evaluation of boar semen. Reproduction in Domestic Animals 40, 392 (2005)
2. Biehl, M., Pasma, P., Pijl, M., Sánchez, L., Petkov, N.: Classification of boar sperm head images using Learning Vector Quantization. In: European Symposium on Artificial Neural Networks (ESANN), pp. 545–550. D-side, Evere (2006)
3. García-Herreros, M., Aparicio, I.M., Barón, F.J., García-Marín, L.J., Gil, M.C.: Standardization of sample preparation, staining and sampling methods for automated sperm head morphometry analysis of boar spermatozoa. International Journal of Andrology 29, 553–563 (2006)
4. Garrett, C., Gordon Baker, H.W.: A new fully automated system for the morphometric analysis of human sperm heads. Fertility and Sterility 63(6), 1306–1317 (1995)
5. Hirai, M., Boersma, A., Hoeflich, A., Wolf, E., Föll, J., Aumüller, T.R., Braun, J.: Objectively measured sperm motility and sperm head morphometry in boars (Sus scrofa): relation to fertility and seminal plasma growth factors. Journal of Andrology 22, 104–110 (2001)
6. Jackson, J.E.: A User's Guide to Principal Components. John Wiley and Sons, Inc., Chichester (1991)
7. Linneberg, C., Salamon, P., Svarer, C., Hansen, L.K.: Towards semen quality assessment using neural networks. In: Neural Networks for Signal Processing IV. Proceedings of the 1994 IEEE Workshop, pp. 509–517 (1994)
8. Alaiz, R., González, M., Alegre, E., Sánchez, L.: Acrosome integrity classification of boar spermatozoon images using dwt and texture techniques. In: International Conference VipIMAGE - I ECCOMAS Thematic Conference on Computational Vision and Medical Image Processing. Taylor and Francis, Abington (2007)
9. Núñez-Martínez, I., Moran, J.M., Peña, F.J.: A three-step statistical procedure to identify sperm kinematic subpopulations in canine ejaculates: Changes after cryopreservation. Reproduction in Domestic Animals 41, 408–415 (2006)

10. Oliva-Hernández, J., Corcuera, B.D., Pérez-Gutiérrez, J.F.: Epidermal Growth Factor (EGF) effects on boar sperm capacitation. Reproduction in Domestic Animals 40, 363 (2005)
11. Otsu, N.: A threshold selection method from gray-level histograms. IEEE Transactions on Systems, Man and Cybernetics 9, 62–66 (1979)
12. Petkov, N., Alegre, E., Biehl, M., Sánchez, L.: LVQ acrosome integrity assessment of boar sperm cells. In: Tavares, J.M.R.S., Natal Jorge, R.M. (eds.) Computational Modelling of Objects Represented in Images - Fundamentals, Methods and Applications; Proc. Int. Symp. CompImage 2006, Coimbra, Portugal, pp. 337–342. Taylor and Francis Group, London (2007)
13. Pinart, E., Bussalleu, E., Yeste, M., Briz, M., Sancho, S., Garcia-Gil, N., Badia, E., Bassols, J., Pruneda, A., Casas, I., Bonet, S.: Assessment of the functional status of boar spermatozoa by multiple staining with fluorochromes. Reproduction in Domestic Animals 40, 356 (2005)
14. Quintero-Moreno, A., Miró, J., Rigau, T., Rodríguez-Gil, J.E.: Regression analyses and motile sperm subpopulation structure study as improving tools in boar semen quality analysis. Theriogenology 61(4), 673–690 (2004)
15. Rozeboom, K.J.: Evaluating boar semen quality. In: Animal Science Facts (2000)
16. Castejón, M., Suarez, S., Alegre, E., Sánchez, L.: Use of statistic texture descriptors to classify boar sperm images applying discriminant analysis. In: International Conference VipIMAGE - I ECCOMAS Thematic Conference on Computational Vision and Medical Image Processing. Taylor and Francis, Abington (2007)
17. Sánchez, L., Petkov, N., Alegre, E.: Classification of boar spermatozoid head images using a model intracellular density distribution. In: Sanfeliu, A., Cortés, M.L. (eds.) CIARP 2005. LNCS, vol. 3773, pp. 154–160. Springer, Heidelberg (2005)
18. Sánchez, L., Petkov, N., Alegre, E.: Statistical approach to boar semen head classification based on intracellular intensity distribution. In: Gagalowicz, A., Philips, W. (eds.) CAIP 2005. LNCS, vol. 3691, pp. 88–95. Springer, Heidelberg (2005)
19. Sánchez, L., Petkov, N., Alegre, E.: Statistical approach to boar semen evaluation using intracellular intensity distribution of head images. Cellular and Molecular Biology 52(6), 53–58 (2006)
20. Thurston, L.M., Watson, P.F., Mileham, A.J., Holt, W.V.: Morphologically distinct sperm subpopulations defined by Fourier shape descriptors in fresh ejaculates correlate with variation in boar semen quality following cryopreservation. Journal of Andrology 22(3), 382–394 (2001)
21. Verstegen, J., Iguer-Ouada, M., Onclin, K.: Computer assisted semen analyzers in andrology research and veterinary practice. Theriogenology 57, 149–179 (2002)
22. Yi, W.J., Park, K.S., Paick, J.S.: Parameterized characterization of elliptic sperm heads using Fourier representation and wavelet transform. In: Proceedings of the 20th Annual International Conference of the IEEE Engineering in Medicine and Biology Society, vol. 20, pp. 974–977 (1998)

HIV-1 Drug Resistance Prediction and Therapy Optimization: A Case Study for the Application of Classification and Clustering Methods

Michal Rosen-Zvi[1], Ehud Aharoni[1], and Joachim Selbig[2]

[1] IBM Research Laboratory in Haifa,
Haifa University, Mount Carmel, Haifa 31905, Israel
rosen@il.ibm.com
[2] University of Potsdam Institute of Biochemistry and Biology,
Max Planck Institute of Molecular Plant Physiology
Am Mühlenberg 1 D-14476 Potsdam-Golm , Germany

Abstract. This chapter provides a review of the challenges machine-learning specialists face when trying to assist virologists by generating an automatic prediction of an outcome of HIV therapy.

Optimizing HIV therapies is crucial since the virus rapidly develops mutations to evade drug pressures. Modern anti-HIV regimens comprise multiple drugs in order to prevent, or at least delay, the development of resistance mutations. In recent years, large databases have been collected to allow the automatic analysis of relations between the virus genome other clinical and demographical information, and the failure or success of a therapy. The EuResist integrated database (EID) collected from about 18500 patients and 65000 different therapies is probably one of the largest clinical genomic databases. Only one third of the therapies in the EID contain therapy response information and only 5% of the therapy records have response information as well as genotypic data. This leads to two specific challenges (a) semi-supervised learning – a setting where many samples are available but only a small proportion of them are labeled and (b) missing data.

We review a novel solution for the first setting: a novel dimensionality reduction framework that binds information theoretic considerations with geometrical constraints over the simplex. The dimensionality reduction framework is formulated to find optimal low dimensional geometric embedding of the simplex that preserves pairwise distances. This novel similarity-based clustering solution was tested on toy data and textual data. We show that this solution, although it outperforms other methods and provides good results on a small sample of the Euresist data, is impractical for the large EuResist dataset. In addition, we review a generative-discriminative prediction system that successfully overcomes the missing value challenge.

Apart from a review of the EuResist project and related challenges, this chapter provides an overview of recent developments in the field of machine learning-based prediction methods for HIV drug resistance.

M. Biehl et al.: (Eds.): Similarity-Based Clustering, LNAI 5400, pp. 185–201, 2009.

1   Introduction

Our world is faced with > 27 million HIV-related deaths to date, and an estimated 4.9 million new infections each year [1]. Highly active antiretroviral therapy (HAART) is the primary approach to treat HIV infection. But often, an accumulation of drug resistant mutations in viral genomes can lead to treatment failure.

Currently, there are several lines of research that attempt to improve HIV diagnostics and therapy based on existing tests and drug compounds. On the experimental side, most activities focus on new methods for the detecting resistance mutations, especially those present as minor populations; this is critical to understanding the mechanisms of multi-drug resistance [2]. Other investigations focus on gaining a better understanding of the host immune system during HIV infection. Schindler et al, [3], showed that the virus-induced down-modulation of a specific host cell gene is an evolutionary highly conserved function of primate lentiviruses and a potential contributor to their nonpathogenic phenotype in most natural hosts. Recently, a large-scale small interfering RNA screen identified more than 250 so-called HIV-depending factors. These host cell proteins participate in a broad array of cellular functions and implicate new pathways in the viral life cycle [4].

Over the last years, several computational methods have been developed whose combined use supports the design of optimized antiretroviral therapies based on several types of data. Around 20 antiretroviral drugs exist today, with most of them are targeted against one out of two proteins essential for the virus replication. There are the protease (PRO) inhibitors and the nucleotide reverse transcriptase (RT) inhibitors. (The RT inhibitors themselves are divided into two types – NNRTI and NRTI.) The computational methods aimed at optimizing the combination of such drugs for a particular patient rely mainly on two components. First, a systematic collection and management of relevant patient data, including sequence information resulting from resistance tests, has to be implemented [5]. Second, sophisticated data analysis methods for analyzing these complex and heterogeneous data have to be provided [6], [7]. Rhee et al. [8] and Carvajal-Rodrigeuez [9] analyzed and compared recent developments in this field. Most of the present-day methods rely on supervised and unsupervised machine learning methods ([10], [11]).

We provide a general overview of HIV interpretation systems, and review those based on machine-learning techniques, in Section 2. Section 3 describes EuResist, a project aimed at developing an interpretation system. We focus on two challenges arising when trying to learn from the large dataset collected in the Euresist project in Section 4 and Section 5. Finally, section 6 includes a summary of the chapter.

2   Machine Learning Interpretation Systems

Several tools exist that aim to predict phenotypic resistance from genotypic data. The majority of available interpretation systems work with hand-crafted tables.

One of the most commonly used tables is the list of mutations maintained by the International Aids Society (IAS). Known as IAS mutations, it is a constantly updated reference list of the PRO and RT sequence (and additional new viral target proteins) mutations that are known to play a role in drug resistance, [12]. Examples of such rule-based interpretation systems are the Agence Nationale de Recherches sur le Sida [13], Rega Institute [14], HIVdb (Stanford University, [15]) and AntiRetroScan (University of Sienna, [16]). They are typically provide a three-level prediction, classifying the virus as resistant, susceptible, or intermediate.

In recent years, several systems were developed that are based on applying machine learning techniques to learn from either in-vitro or clinical-genomic data. One such system using in-vitro data is the Geno2pheno system, which uses decision trees, information theoretic analysis, and Support Vector Regression (SVR) to provide a full range predictions [17,10]. Rhee et al. [8] examine five different statistical learning techniques, namely decision trees, neural networks, SVR, least square regression, and least angle regression for the prediction task. All these methods were found to have similar performance.

Committees of Artificial Neural Networks have been used to predict the change in VL after treatment start, along with information about the viral sequences and the intended regimen, CD4+ counts, baseline VL, and four treatment history indicators that are used as input features[18]. Geno2pheno-THEO by [19], another online tool, predicts the success probability of a putative treatment, based on sequence information and the estimated genetic barrier of the viral variant to resistance against every drug in the combination therapy. However, Geno2Pheno-THEO does not make use of information about the patient or other available clinical markers. Saigo et al., [20], propose a nonlinear regression-based method, called itemset boosting, in the space of power sets of mutations in genotypes. The method constructs a new feature space by progressively adding a complex feature in each iteration. New features are selected in order to minimize the gap between current predictions and given target values.

Several attempts have been undertaken to develop computational methods for molecular modeling of HIV-1 enzyme-inhibitor complexes and to build Qualitative Structure-Activity Relationship (QSAR) models for inhibitor activity prediction. However, these methods have not been applied to predict resistance. Almerico et al. [21] propose an approach to evaluate the features of inhibitors that are less likely to trigger resistance or are effective against mutant HIV strains, on the basis of physico-chemical descriptors and structural similarity. The problem with these type of approaches is, that in most cases, only a few data points (compounds) are available described by a large number of heterogeneous features. They applied Principal Component Analysis (PCA) combined with Discriminant Analysis (DA) to explore the activity of inhibitors against HIV-1 reverse transcriptase and protease, using data from docking these inhibitors into the wild type molecules as well as into structures modified due to mutations. QSAR investigations may also help to classify new inhibitors, if the structural descriptors are able to reflect functional properties. A critical point in the context of designing experiments related to new drugs is the similarity measure used to compare drug-like molecules. Classification experiments based on topological features only failed to separate inhibitors from noninhibitors of HIV-1 protease, cf. [6]. These limitations may be overcome by considering conformational ensembles and applying sophisticated clustering methods in order to identify cluster representatives.

The EuResist system[1] is a system being developed and designed to provide a recommendation for a HAART therapy, given the genotype [22,23]. The recommendation is expected to improve as more features such as history therapies, baseline viral load, and so on are provided. The system scans through known combinations and provides the most promising therapies based on the likelihood of success as derived by an integrated system that combines three engines, each based on a different learning technology. In the next section we provide details on the development of this system and in Section 5 we provide details regarding one of the core technologies.

3   The EuResist Project

The EuResist project is a European Union funded project, aimed at developing a European integrated system for clinical management of antiretroviral drug resistance [24]. The system provides clinicians with a prediction of response to antiretroviral treatment in HIV patients, thus helping clinicians choose the best drugs and drug combinations for any given HIV genetic variant. To this end a huge database was created, aggregating the clinical records of HIV patients from various sources: the ARCA database (Italy)[2], AREVIR database (Germany) [5], data coming from the Karolinska Infectious Diseases and Clinical Virology Department (Sweden), and a smaller data set from Luxembourg. Several prediction engines were trained using this data and their predictions is be freely available via a web interface.

In what follows, we describe the definitions designed by HIV specialists to enable the automatic analysis of the integrated database (IDB) [25]. The basic data item, a *therapy*, is defined. Each therapy describes a period of time in which a certain HIV patient was treated with a specific treatment. This description includes: the relevant HIV gene sequences identified in this patient's blood prior to therapy start, a viral load measure (the count of HIV in the patient's blood sample) prior to therapy start, the drug compounds used in this therapy, and a viral load measure taken four to twelve weeks after therapy start, indicating the efficiency of this therapy. If this second viral load measure result is "undetectable", meaning the amount of HIV has dropped below the detection threshold, (500 copies per milliliter), the therapy is considered a success. It is also considered a success if the second viral load measure has dropped by 2 points on a logarithmic scale compared to the initial viral load measure. In any other case a therapy is considered a failure.

---

[1] http://engine.euresist.org
[2] http://www.hivarca.net/

Some other features are available in the IDB, such as the patient's age or country of origin, but by far the most informative features are the HIV gene sequences and the drug compounds used. The appearance of certain combinations of mutations in a sequence gives the virus full or partial resistance to certain drug compounds. Thus, the main goal when designing and training the prediction engines was to correctly capture this relationship.

Approximately 3000 therapies were available for training the prediction engines. These therapies, however, are only a fraction of the data collected in the EuResist database. The remainder of the therapies (approximately 60000), contain insufficient data for use as training data. In particular, there are numerous therapies (approximately 10000) where the therapy data contained the HIV gene sequences but was missing one of the viral load measures. There are approximate 17000 therapies with sufficient viral load measures but no genotype information.

The classical approach is to ignore the incomplete data and train and test on the complete data. The results of such analysis are provided in Section 2. In what follows we review two methods that make use of the incomplete data records.

4   A Detour - Semi-supervised Learning

Learning from genotype data, or as it is in the case of the IDB learning from two the RT and PRO proteins related to the replication of the virus, is challenging as the number of amino acids in the protein sequence is large and not all positions are relevant to the analysis. Sequences often contain uncertain information due to inaccurate sequencing, e.g., positions where one of several mutations might have occurred. Thus we chose to represent them using probability distributions. To meet all these challenges, the IBpairs has been developed[3] specifically for data composed of probability distributions. In what follows we review this dimensionality reduction approach. The approach was found to be successful on small data but not feasible for large datasets.

This study addresses the setting of semi-supervised learning, under which many samples are available, but only a small fraction of them are labeled. Moreover, our samples lie in a high dimensional simplex; namely, each sample is in fact a conditional probability distribution. This setting is common in real world applications in which measurements are probabilistic, rather than deterministic, and the cost of the labels is high.

The following sections review the IBpairs dimensionality reduction rationale, which binds information considerations with geometrical constraints over the simplex. This rationale was formulated as an optimization problem. We show that the optimal solution is approximated by the solution of the Information Bottleneck criterion. This provides a novel geometrical view of the Information Bottleneck, which allows us to link it with a suitable kernel, leveraging it from a dimension-reduction scheme to a self-contained semi-supervised approach. We further demonstrate this approach on synthetic and text categorization data.

---

[3] Developed by EH and MRZ, previously presented shortly at [26].

4.1   Preliminaries and Background

Let X and Y be two discrete random variables that take values in the sets $\mathcal{X}$ and $\mathcal{Y}$, respectively. A set of samples is provided in which each sample represents a conditional probability distribution $P(x|y)$. This can be given either explicitly or implicitly through many i.i.d samples drawn from the joint distribution of X and Y. When given explicitly, the values of Y can be regarded as the samples' indices. In a simple example, X represents words and Y documents.

Typically, for the semi-supervised setting, few of these samples are associated with labels in $Z = \{0, 1\}$, representing, for example, documents' categories. The goal is to predict the label z_s of a previously unlabeled sample s, given its features vector $P(X|Y = s)$. A common approach to semi-supervised learning is to employ a dual-phase algorithm, first to apply a dimensionality reduction procedure that extracts information from the large number of unlabeled examples, assuming that the many features are redundant in some way or from computational considerations. The second phase contains a discriminative algorithm that performs the supervised learning in the lower dimension space. In the proposed approach, the second phase contains a stat-of-the-art discriminative algorithm support-vector machine, where the main issue is to select the appropriate kernel mapping. Hence, in the first phase the variable X is replaced by a simpler variable $\hat{X}$ of lower cardinality, $|\hat{X}| < |X|$. The challenge is to find an effective transformation to the simpler variable such that the information relevant to the supervised learning is not lost.

A host of information-theoretic divergence measures between two probability distributions has been extensively studied over the years [27,28]. Among these, the Jensen-Shannon (JS) divergence is widely used. This measure is identical to the mutual information between X and Y. Sanov's theorem proves that this measure quantifies the difficulty of distinguishing between two distributions in a statistical test [27]. It is nonnegative, symmetric, and vanishes if and only if the two probability distributions are identical. However, it is not a metric, as the triangle inequality does not prevail. Nevertheless, Endres and Schindelin [29] recently introduced a new metric space for probability distributions. Their metric space (Ω, M_{PQ}^{JS}) is defined over $\Omega = \{p \in \mathcal{R}_+^n \| \sum_{i=1}^n p_i = 1\}$ where $\mathcal{R}_+^n$ is the positive n-dimensional real space including zero and $M^{JS}(P, Q) = \sqrt{\sum_{i=1}^n \left(p_i \ln \frac{2p_i}{p_i + q_i} + q_i \ln \frac{2q_i}{p_i + q_i} \right)}$. We call this metric, which is simply the square-root of the JS divergence, the JS metric.

A natural way to design a similarity mapping out of the metric is to derive it from the natural norms (that is, the distance of the point X from some zero point $|X| = M_{XX_0}^{JS}$), $K(x, y) = (|X + Y|^2 - |X|^2 - |Y|^2)/2$. No matter what the zero point is, the first term in the left-hand side defines a point that does not exist in the metric (outside the simplex) and thus this mapping cannot be used. Another alternative is to define the mapping $K(x_1, x_2) = \exp(-D(x_1, x_2)^2)$, which has the same form as RBF kernels in the Euclidean space and is a metric kernel [30]. We focus on this JS kernel as it is consistent with our dimensionality reduction scheme. A similar kernel is the diffusion kernel [31], which is obtained by deriving the geodesic distance in the Fisher metric.

4.2 Problem Formulation

Definition 1: Pair information. Let X and Y be two discrete random variables, $P(X|Y)$, the conditional probability distribution of X given Y, and $P(Y)$, some prior over Y. For each pair of values $\{i, j\}$ that Y can take, we define the *pair probability* as the probability of one of these values $y \in \{i, j\}$ when the sample space is limited to this specific pair, namely, $P^{ij}(y) = P(Y = y|Y = \{i, j\}) = \frac{P(Y=y)}{P(Y=i)+P(Y=j)}$ where the convention is that there are no zero probabilities, $\forall y, \ P(Y = y) \neq 0$.

The *pair joint distribution* is defined as $P^{ij}(x, y) = P^{ij}(y)P(x|y)$, the *marginal* is defined as $P^{ij}(x) = P(x|Y = \{i, j\}) = \sum_{y\in\{i,j\}} P^{ij}(y)P(x|y)$ and the definition of the *pair information* follows

$$I^{ij}(X;Y) = \sum_{x\in\mathcal{X}, y\in\{i,j\}} P(x|y)P^{ij}(y) \log \frac{P(x|y)}{P^{ij}(x)} \tag{1}$$

Note that for a uniformly distributed variable Y (with $P(Y) = 1/|Y|$ and any conditional distribution $P(X|Y)$) the pair information is identical to the square of the distance between a pair of probabilities in the JS metric, i.e.,

$$M^{JS}\left(P(X|Y = i), P(X|Y = j)\right) = \sqrt{2I^{ij}(X;Y)} \tag{2}$$

Dimensionality reduction criterion. Assuming that the distances between instances, measured by the JS metric are relevant for supervised learning, we attempt to find a mapping $P(\hat{X}|X)$ from X to $\hat{X}$ such that irrelevant information on X is lost while the distances between the original sample points are distorted as little as possible. The square of the distance between each pair of points in the original space equals (up to a factor of two) the pair information, $I^{ij}(X, Y)$. Each of the square distances between pairs of points in the original dimension upper bounds the corresponding square distance in the lower dimension space, $\left[M^{JS}(P(\hat{X}|Y = i); P(\hat{X}|Y = i))\right]^2$ due to the data processing inequality [27]. Thus, it is enough to require that the distances be lower bounded by some constant M_D. In principle, there should be $|Y|(|Y| - 1)/2$ constraints, each lower bounds each pair of instances. Such a requirement would result in an intractable algorithm and thus we replace the set of constraints, by a single constraint over the averaged distortion,

$$\min_{P(\hat{X}|X)\in\mathcal{F}} I(\hat{X}; X) \tag{3}$$

where $\mathcal{F}$ is the set of probability distributions $P(\hat{X}|X)$ s.t. $\frac{2}{|Y|(|Y|-1)} \sum_{\{i,j\}\in Y} \left[M^{JS}(P(\hat{X}|i); P(\hat{X}|j))\right]^2 > M_D$. In the notation of pair information, and for uniformly distributed Y, this constraint is equivalent to $\frac{4}{|Y|(|Y|-1)} \sum_{\{i,j\}\in Y} I^{ij}(\hat{X}; Y) > M_D$.

The optimization problem formalized above can be solved directly. The criteria shown in equation 3 is similar to that defined by the Information Bottleneck (IB) optimization problem[32]. Therefore, we chose to optimize equation 3 using an algorithm built along the same lines as the one for IB, which we term IBPairs. We do not describe it here for lack of space.

4.3   Experimental Results

The Polya Urn Toy Model. We employed the Polya urn process to generate noisy copies of multinomial distributions. In the Polya urn process, multinomial distributions are represented by marbles in an urn. Assume there is an urn with t marbles each colored with one out of $|X|$ possible colors. The probability distribution that by picking one marble at random from the urn its color would be the xth color is $P(X = x) = \frac{n_x}{\sum_x n_x}$ where n_x stands for the number of x-colored marbles in the urn. According to the Polya urn scheme, in each iteration a marble is picked at random from the urn, its color is inspected, and then the marble is returned to the urn along with an identical (color-wise) new marble. Thus an urn that initially contains T marbles, would contain after t iterations $t + T$ marbles. Furthermore, if we have two identical urns with K marbles and we perform the iterations on each one separately, the more iterations are carried out the more the distributions represented by the urn are different . It is proved that at the limit $t \rightarrow \infty$ $P(X)$ is uniformly distributed over the simplex (see e.g., [33]). In other words, if we start with two identical urns, after a large enough number of iterations, it would be impossible to know whether there is a mutual initial archetype for both urns.

Data generation and learning. To generate samples from the Polya urn model, two initial archetype distributions were selected, each generated by $T = 43$ marbles of $|X| = 40$ different colors. These archetypes define the classification of the data. One hundred samples were generated from each archetype by repeatedly starting from it and iterating $t = 1100$ times. A balanced set of 200 samples was selected for the training stage, with only 50 of them labeled. First, the dimension reduction procedure (either IB or IBPairs) was executed, using the 200 samples. Then SVM was trained in 5-fold cross validation over the 50 labeled samples. This procedure was repeated with several values of SVM cost and β, to select an optimal value. The results present the mean error rate and standard deviation achieved repeating the experiments 10 times with different sets of labeled samples randomly selected from the 200.

Kernels Comparison. This experiment compares the generalization performance of the different kernels with the IB. The test set is constructed of the 2000 samples that were never used for training. Table 1 shows that the best performance is achieved when using the IB with the JS kernel. The superiority of the JS kernel is also visualized in Figure 1. Each dot in the figure compares the error rate achieved with the JS kernel, in comparison to the other kernel type.

Table 1. Generalization setting: error rate of different kernels on the Polya urn data compressed with IB

Kernel Type	Error Rate
Linear	0.21± 0.06
RBF	0.22± 0.06
JS	0.18± 0.01

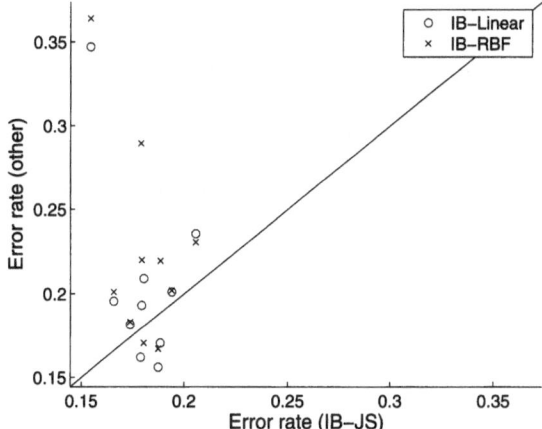

Fig. 1. Generalization setting: a scatter plot of error rate of Linear and RBF kernels on the Polya urn data compressed with IB versus the error rate of the JS kernel

Transductive learning. To compare our scheme to other algorithms we executed this procedure in a transductive mode, with the test set being the unlabeled 150 samples out of the 200 used for the dimension reduction. Figure 2 presents the error rates and standard deviations achieved by each algorithm. The highest error rate was achieved by the naive classification, using the nearest neighbor with Euclidean distances. The SVM with a linear kernel and no dimension reduction scheme shows some improvement. In both cases, the error rate is reduced by simply converting to the JS metric, demonstrating our claim that this is the most suitable metric for this type of data (second and 5th bars). The fourth bar presents the results of the Low Density Separation (LDS) algorithm, considered state-of-the-art in transductive applications. This algorithm is based on transductive SVM with a modified objective function that places the decision boundary in low density regions. The selection of the parameters for the LDS was performed as described in [34]. The added value of employing an IB/IBPairs compression algorithm with the JS kernel is apparent from the 7th (last) bar. Note that both the IB-JS and the IBPairs-JS algorithms yielded almost identical errors and thus only the IB-JS result is presented here.

Figure 3 compares the IB-JS approach with the LDS. Every circle in the scatter plot represents the error rate on a single test set. Each test set is a

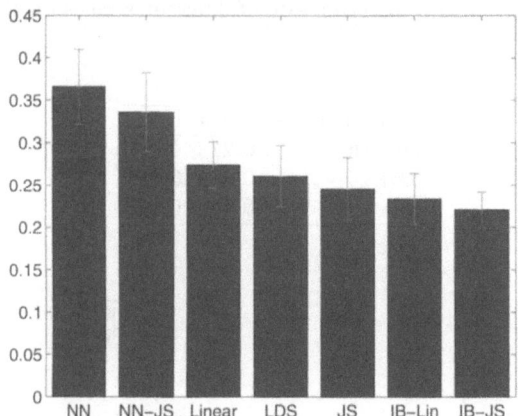

Fig. 2. The performance of different transductive algorithms on the Polya urn data

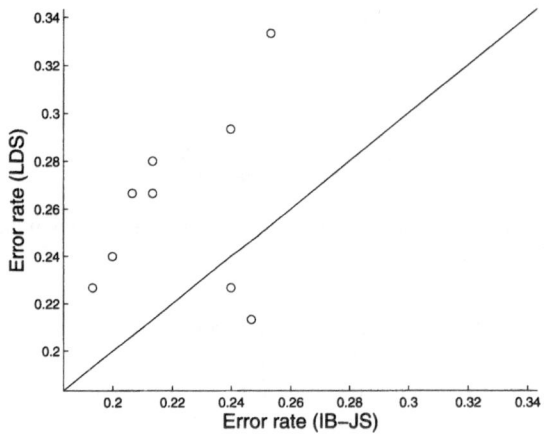

Fig. 3. Transductive setting: a comparison of the error rates of LDS vs. IB-JS on 10 randomly labeled sets of the Polya urn data

randomly selected labeled set of 50 samples out of the 200. The associated p-value obtained using a t-test over these tests is 0.008.

Text data. We tested the algorithm on a subset of the 20newsgroup using 400 documents from 2 groups out of the 4 and with the 100 selected words[4]. We applied the IB-JS and the LDS algorithms in a transductive setting with the same model selection scheme as above. In each of the 10 partitions, the label of 100 randomly selected documents is observed and 300 documents serve as unlabeled data. Both algorithms achieved similar accuracy of 0.08 ± 0.02 error rate. Note, the dimensionality reduction part of the IB-JS algorithm can provide

---

[4] Taken from http://www.cs.toronto.edu/~roweis/data.html

insights regarding the different clusters as opposed to the LDS algorithm. In Figure 4) we present the five most likely words per the five clusters derived by the IB algorithms. The edges in the graph indicate the small JS distance between clusters. Indeed the five clusters are divided into two separate categories, *computers* and *politics*, which are the original two selected groups.

4.4 Discussion and Conclusion

In this section we described a semi-supervised scheme that is suited to fit probabilistic data. This type of data is very common in real world applications in which measurements may be stochastic and labels are costly.

The design of this scheme was driven by the needs of the EuResist project, as described in section 3. We managed to successfully apply this scheme when experimenting with a small number of samples (80 labeled and 4000 unlabeled samples)., as shown in Figure 5. We used an SVM classifier to predict whether a therapy would be a success or failure. The X axis shows the amount of labeled samples used for training The Y axis shows the precision achieved on the rest of the 80 labeled samples. We compared runs with various kernels: linear, RBF and JS (Distr), with and without a prior stage of compression using IBPairs. The results show that IBPairs works well with the JS kernel, as anticipated. It further shows that this approach maximizes the results and surpasses previous known techniques that use regression based on invitro phenotypic tests (making use of SVM, see [17]), marked by the horizontal black line in the graph.

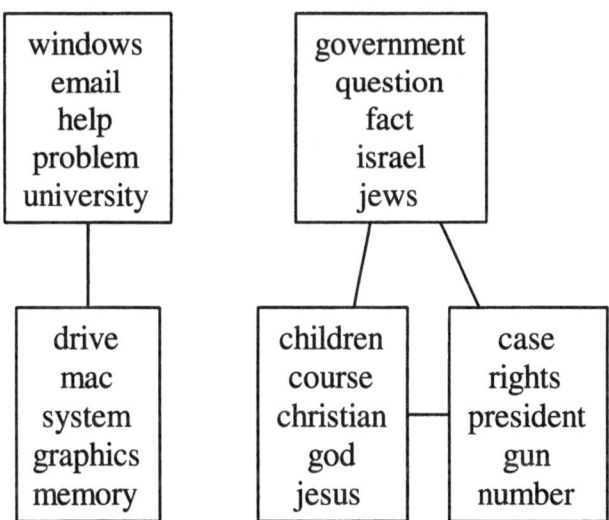

Fig. 4. First highest ranked words from every cluster

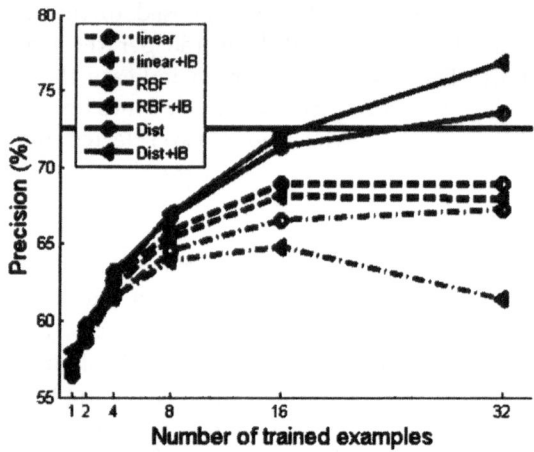

Fig. 5. Results of various schemes applied to early versions of the EuResist database containing 80 labeled and 4000 unlabeled samples

5   A Generative-Discriminative Approach

This sections describes an engine trained on the EuResist dataset by applying generative models of the interactions between current and history antiretroviral drugs. It is well known that subsequent use of previously employed and closely related drugs may be inefficient (see e.g., [35,36]). Generative models are powerful when some expert knowledge for guidance of the design of the network is available ([37]). In this case, we have 20 drugs naturally divided into three classes: PI, NRTI, and NNRTI. A number of candidate networks containing nodes representing the drugs individually and/or the drug classes are tested using 10-fold cross-validation with 20 divisions on a large training set. The data available for training is the incomplete labeled set along with the training dataset, for a total of about 20000 therapies. The network selected is shown in Figure 6. The root node stands for success or failure and the leaves are binary nodes standing for the existence/not existence of a drug in current therapy. The three nodes in the middle are discrete nodes that stand for the number of history drugs adhered by the patient, with a separate count per drug class. Each of these three nodes is a parent to the drugs from the related class. This Bayesian network, with no other features like genotype, yields an AUC of 0.716 ± 0.001. It outperforms a Naive Bayes network trained with 6 leaves - count for history and current drugs (AUC of 0.698 ± 0.002) and many other alternative networks. A very similar performance is obtained with Naive Bayes trained with the same set of nodes. With 23 leaves, 3 stand for the count of history drugs per group and 20 stand for the existence /not existence of each drug in the current therapy (AUC of 0.715 ± 0.001). The contribution of other demographical features, like age, to the prediction performance of the network were also tested but found negligible.

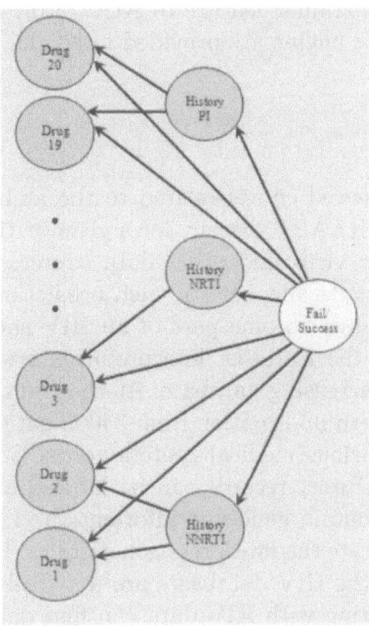

Fig. 6. Bayesian network used in the generative-discriminative engine

This core generative engine generates a prediction of success/failure and a number in the range [0 1], which is used as a training feature. On top of this core, a second layer of discriminative engine is trained. The second layer engine uses logistic regression and all candidate features from the data are tested for their contribution to this second layer. The features selected to stand for the genotype mutations are mutations with a high correlation of success/failure, 20/25 mutations of PRO/RT in the baseline genotype, and 5 mutations appearing in history genotype. (See the Appendix for more details.) The final list of features used in the generative discriminative engine contains, in addition to the above features, the 20 compounds, baseline viral load, and the number of past treatment lines.

In a similar process, a generative discriminative engine was developed for the case of the minimal feature set. The best performing Bayesian network is similar to the one in Figure 6, except that the middle layer with history drugs is replaced by an indicator that is 0/1 if the drug belongs/does not belong to a drug class. This Bayesian network alone provides results with an AUC of 0.672 ± 0.001. It outperforms a Naive Bayes network trained with 20 leaves that stand for the drugs (AUC of 0.654 ± 0.001) and a logistic regression engine trained on the same features (AUC of 0.662 ± 0.001). The top layer of logistic regression engine is trained on clinical genomic data with the 20 drugs and 20/25 mutations selected based on correlation with viral load drop of PRO/RT, results in AUC of 0.744 ± 0.003. This performance improves when combined with the

single Bayesian network trained feature to AUC of 0.749 ± 0.003. More results
obtained by training this engine are provided in [22,23].

6 Summary

In this chapter, we reviewed efforts related to the ambitious challenge of predicting the success of a HAART therapy for a given patient, based on the genotype information of the virus and other data sources about the patient and
the virus. Over the past decade, several such prediction systems have become
available and seem to have become part of an HIV specialist's common practice. We elaborated on the EuResist interpretation system, which is based on
mining a dynamically increasing database. At the end of 2008, this database is
the largest worldwide with information from 34000 patients, treated with 98000
different therapies in various medical centers across Europe. Around 32000 of
the therapies in the EuResist records can be labeled as success or failure and
5000 of the therapies contain genotype information. The larger the volume of
the data, the more accurate the interpretation systems. Due to the nature of this
personal data, most of the HIV databases are not freely available. Researchers
interested in experimenting with HIV data can find data available at Stanford
database[5].

More generally, this review corroborates the personalized treatment approach
where therapy provided to a patient can be optimized based on the individual's
personal clinical-genomic factors. This optimized selection needs to be based
on large scale clinical genomic databases enhanced with analytical tools. The
advanced machine-learning techniques available today, along with the abundance
of medical data, can provide a powerful decision support system and contribute
to the emerging area of information based medicine.

Acknowledgements

MRZ and EH gratefully acknowledge the EuResist funding and the EuREsist
partners for this work (EU FP6 IST-2004-027173). JS and MRZ are thankful for
being invited to attend a Dagstuhl seminar on *Similarity-based Clustering and
its Application to Medicine and Biology* where part of this work was presented.

References

1. Altfeld, M., Alle, T.: Hitting hiv where it hurts: an alternative approach to hiv
   vaccine design. TRENDS in Immunology 27, 504–510 (2006)
2. Cai, F., Haifeng, C., Hicks, C., Bartlett, J., Zhu, J., Gao, F.: Detection of minor
   drug-resistant populations by parallel allele-specific sequencing. Nature Methods 4,
   123–125 (2007)

---

[5] http://hivdb.stanford.edu/

3. Schindler, M., Mönch, J., Kutsch, O., Li, H., Santiago, M.L., Billet-Ruche, F., Müller-Trutwein, M.C., Novembre, F.J., Peeters, M., Courgnaud, V., Bailes, E., Roques, P., Sodora, D.L., Silvetri, G., Sharp, P.M., Hahn, B.H., Kirchhoff, F.: Nef-mediated suppression of t cell activation was lost in a lentiviral lineage that gave rise to hiv-1. Cell 125, 1055–1067 (2006)
4. Brass, A.L., Dykxhoorn, D.M., Benita, Y., Yan, N., Engelman, A., Xavier, R.J., Lieberman, J., Elledge, S.J.: Identification of host proteins required for hiv infection through a functional genomic screen. Science, 1152725 (January 2008)
5. Roomp, K., Beerenwinkel, N., Sing, T., Schülter, E., Büch, J., Sierra-Aragon, S., Däumer, M., Hoffmann, D., Kaiser, R., Lengauer, T., Selbig, J.: Arevir: A secure platform for designing personalized antiretroviral therapies against hiv. In: Leser, U., Naumann, F., Eckman, B. (eds.) DILS 2006. LNCS (LNBI), vol. 4075, pp. 185–194. Springer, Heidelberg (2006)
6. Cordes, F., Kaiser, R., Selbig, J.: Bioinformatics approach to predicting hiv drug resistance. Expert Review of Molecular Diagnostics 6(2), 207–215 (2006)
7. Altmann, A., Rosen-Zvi, M., Prosperi, M., Aharoni, E., Neuvirth, H., Schülter, E., Büch, J., Peres, Y., Incardona, F., Sönnerborg, A., Kaiser, R., Zazzi, M., Lengauer, T.: The euresist approach for predicting response to anti hiv-1 therapy. In: The 6th European HIV Drug Resistance Workshop, Cascais, Portugal (2008)
8. Rhee, S.Y., Taylor, J., Wadhera, G., Ben-Hur, A., Brutlag, B., Shafer, R.: Genotypic predictors of human immunodeficiency virus type 1 drug resistance. PNAS 103, 17355–17360 (2006)
9. Carvajal-Rodriguez, A.: The importance of bio-computational tools for predicting hiv drug re-sistance. Cell 1, 63–68 (2007)
10. Beerenwinkel, N., Schmidt, B., Walther, H., Kaiser, R., Lengauer, T., Hoffmann, R., Korn, K., Selbig, J.: Diversity and complexity of hiv-1 drug resitance: A bioinformatics approach to predicting phenotype from genotype. PNAS 99, 8271–8276 (2002)
11. Altmann, A., Beerenwinkel, N., Sing, T., Savenkov, I., Doumer, M., Kaiser, R., Rhee, S.Y., Fessel, W., Shafer, R., Lengauer, T.: Super learning: A applica-tion to prediction of hiv-1 drug resistance. Statistical Applications in Genetics and Molecular Biology 6(7), 169–178 (2007)
12. Johnson, A., Brun-Vezinet, F., Clotet, B., Gunthard, H., Kuritzkes, D., Pillay, D., Schapiro, J., Richman, D.: Update of the drug resistance mutations in hiv-1: 2007. Top HIV Med. 15(4), 119–125 (1991)
13. Brun-Vezinet, F., Costagliola, D., Mounir Ait, K., Calvez, V., Clavel, F., Clotet, B., Haubrich, R., Kempf, D., King, M., Kuritzkes, D., Lanier, R., Miller, M., Miller, V., Phillips, A., Pillay, D., Schapiro, J., Scott, J., Shafer, R., Zazzi, M., Zolopa, A., DeGruttola, V.: Clinically validated genotype analysis: guiding principles and statistical concerns. Antiviral therapy 9(4), 465–478 (2004)
14. Van Laethem, K., De Luca, A., Antinori, A., Cingolani, A., Perna, C., Vandamme, A.: A genotypic drug resistance interpretation algorithm that significantly predicts therapy response in hiv-1-infected patients. Antiviral therapy 2, 123–129 (2002)
15. Kantor, R., Machekano, R., Gonzales, M.J., Dupnik, K., Schapiro, J.M., Shafer, R.W.: Human immunodeficiency virus reverse transcriptase and protease sequence database: an expanded data model integrating natural language text and sequence analysis programs. Nucleic Acids Research 29(1), 296–299 (2001)
16. Zazzi, M., Romano, L., Venturi, G., Shafer, R., Reid, C., Dal Bello, F., Parolin, C., Palu, G., Valensin, P.: Comparative evaluation of three computerized algorithms for prediction of antiretroviral susceptibility from hiv type 1 genotype. J. Antimicrob Chemother 53(2), 356–360 (2004)

17. Beerenwinkel, N., Däumer, M., Oette, M., Korn, K., Hoffmann, D., Kaiser, R., Lengauer, T., Selbig, J., Walter, H.: Geno2pheno: estimating phenotypic drug resistance from hiv-1 genotypes. Nucleic Acids Research 31(13), 3850–3855 (2003)
18. Larder, B., Wang, D., Revell, A., Montaner, J., Harrigan, R., De Wolf, F., Lange, J., Wegner, S., Ruiz, L., Perez-Elias, M., Emery, S., Gatell, J., Monforte, A., Torti, C., Zazzi, M., Lane, L.: The development of artificial neural networks to predict virological response to combination hiv therapy. Antiviral therapy 12, 15–24 (2007)
19. Altmann, A., Beerenwinkel, N., Sing, T., Savenkov, I., Doumer, M., Kaiser, R., Rhee, S.Y., Fessel, W., Shafer, R., Lengauer, T.: Improved prediction of response to antiretroviral combination therapy using the genetic barrier to drug resistance. Antiviral therapy 12, 169–178 (2007)
20. Saigo, H., Uno, T., Tsuda, K.: Mining complex genotypic features for predicting hiv-1 drug resistance. Bioinformatics 23(18), 2455–2462 (2007)
21. Almerico, A., Tutone, M., Lauria, A.: Docking and multivariate methods to explore hiv-1 drug-resistance: a comparative analysis. J. Comput. Aided Mol. Des. (2008)
22. Altmann, A., Rosen-Zvi, M., Prosperi, M., Aharoni, E., Neuvirth, H., Schülter, E., Büch, J., Struck, D., Peres, Y., Incardona, F., Sönnerborg, A., Kaiser, R., Zazzi, M., Lengauer, T.: Comparison of classifier fusion methods for predicting response to anti hiv-1 therapy. PLoS ONE 3(10), 3470 (2008)
23. Rosen-Zvi, M., Altmann, A., Prosperi, M., Aharoni, E., Neuvirth, H., Sönnerborg, A., Schülter, E., Struck, D., Peres, Y., Incardona, F., Kaiser, R., Zazzi, M., Lengauer, T.: Selecting anti-HIV therapies based on a variety of genomic and clinical factors. Bioinformatics 24(13), i399–i406 (2008)
24. Aharoni, E., Altman, A., Borgulya, G., D'Autilia, R., Incardona, F., Kaiser, R., Kent, C., Lengauer, T., Neuvirth, H., Peres, Y., Petroczi, A., Prosperi, M., Rosen-Zvi, M., Schülter, E., Sing, T., Sönnenborg, A., Thompson, R., Zazzi, M.: Integration of viral genomics with clinical data to predict response to anti-hiv treatment. In: IST-Africa 2007 Conference & Exhibition (2007)
25. Zazzi, M., Aharoni, E., Altmann, A., Baszó, F., Bidgood, P., Borgulya, G., Denholm-Prince, J., Fielder, M., Kent, C., Lengauer, T., Nepusz, T., Neuvirth, H., Peres, Y., Petroczi, A., Prosperi, M., Romano, L., Rosen-Zvi, M., Schülter, E., Sing, T., Sönnerborg, A., Thompson, R., Ulivi, G., Zalány, L., Incardona, F.: Euresist: exploration of multiple modeling techniques for prediction of response to treatment. In: Proceedings of the 5th European HIV Drug Resistance Workshop (2007)
26. Rosen-Zvi, M., Neuvirth, H., Aharoni, E., Zazzi, M., Tishby, N.: Consistent dimensionality reduction scheme and its application to clinical hiv data. In: NIPS 2006 workshop, Novel Applications of Dimensionality Reduction (2006)
27. Cover, T.M., Thomas, J.A.: Elements of information theory. Wiley-Interscience, New York (1991)
28. Lin, J.: Divergence measures based on the shannon entropy. IEEE Transactions on Information Theory 37(1), 145–151 (1991)
29. Endres, D.M., Schindelin, J.E.: A new metric for probability distributions. IEEE Transactions on Information Theory 49(7), 1858–1860 (2003)
30. Chan, A.B., Chan, A.B., Vasconcelos, N., Vasconcelos, N., Moreno, P.J., Moreno, P.J.: A family of probabilistic kernels based on information divergence (2004)
31. Lafferty, J., Lebanon, G.: Diffusion kernels on statistical manifolds. J. Mach. Learn. Res. 6, 129–163 (2005)
32. Tishby, N., Pereira, F., Bialek, W.: The information bottleneck method. In: Proceedings of the 37th Annual Allerton Conference on Communication, Control and Computing, pp. 368–377 (1999)

33. Chung, F., Handjani, S., Jungreis, D.: Generalizations of polya's urn problem. Annals of combinatorics 7, 141–154 (2003)
34. Chapelle, O., Zien, A.: Semi-supervised classification by low density separation. In: Cowell, R., Ghahramani, Z. (eds.) Tenth International Workshop on Artificial Intelligence and Statistics, pp. 57–64 (2005)
35. Siliciano, R.: Viral reservoirs and ongoing virus replication in patients on haart: implications for clinical management. In: Conf. Retrovir Oppor. Infect Conf. Retrovir Oppor. Infect 8th Abstract No. L5 (2001)
36. Piliero, P.: Early factors in successful anti-hiv treatment. Journal of the International Association of Physicians in AIDS Care (JIAPAC) 2(1), 10–20 (2003)
37. Pearl, J.: Probabilistic Reasoning in Intelligent Systems: Networks of Plausible Inference. Morgan Kaufmann, Santa Mateo (1988)

Author Index